Oscillations

Theory and applications in AFM

Online at: https://doi.org/10.1088/978-0-7503-5809-5

Oscillations

Theory and applications in AFM

Tuza Adeyemi Olukan
ARC-Arctic Centre for Sustainable Energy, Department of Physics and Technology, UiT The Arctic University of Norway, 9010 Tromsø, Norway

Sergio Santos
ARC-Arctic Centre for Sustainable Energy, Department of Physics and Technology, UiT The Arctic University of Norway, 9010 Tromsø, Norway

Lamiaa Sami Elsherbiny
Laboratory for Energy and Nano Science, Masdar Campus, Khalifa University, Abu Dhabi, United Arab Emirates

Matteo Chiesa
ARC-Arctic Centre for Sustainable Energy, Department of Physics and Technology, UiT The Arctic University of Norway, 9010 Tromsø, Norway
and
Laboratory for Energy and Nano Science, Masdar Campus, Khalifa University, Abu Dhabi, United Arab Emirates

IOP Publishing, Bristol, UK

ISBN 978-0-7503-5809-5 (ebook)
ISBN 978-0-7503-5807-1 (print)
ISBN 978-0-7503-5810-1 (myPrint)
ISBN 978-0-7503-5808-8 (mobi)

DOI 10.1088/978-0-7503-5809-5

Version: 20240901

IOP ebooks

British Library Cataloguing-in-Publication Data: A catalogue record for this book is available from the British Library.

Published by IOP Publishing, wholly owned by The Institute of Physics, London

IOP Publishing, No.2 The Distillery, Glassfields, Avon Street, Bristol, BS2 0GR, UK

US Office: IOP Publishing, Inc., 190 North Independence Mall West, Suite 601, Philadelphia, PA 19106, USA

Many great alumni and professors have contributed to the LENS lab for almost two decades. Throughout these years we have held many group meetings with presentations and great discussions following them, attended conferences, thought of experiments together and written many exciting articles. In this text we have tried to condense all the knowledge accumulated over these years on oscillations and vibrations theory in the lab with particular attention to applications in atomic force microscopy. For all the above the authors thank all of these great contributors and hope we can continue exploring new avenues in science and technology together in the future.

Contents

Author biographies

Tuza Olukan

Tuza Olukan is a Postdoctoral Research Fellow at UiT, Norway. He holds a doctorate degree in Material Science and Nanotechnology Engineering under the MIT/MI Collaborative Program at the Khalifa University. His scientific interests rest on the deployment of cutting-edge solutions in addressing scientific and technological challenges with an emphasis on low-cost, sustainable, and, locally sourced materials and resources. His current research effort is geared towards adopting multidisciplinary strategies in the decarbonization, decentralization, and digitalization of the electricity sector especially in remote regions of the world and the Global South countries in general.

Sergio Santos

Sergio Santos is an author in science and law. When the theme for the book was first conceived, he was a Postdoctoral Research Fellow at UiT (2019-2022), Norway. S Santos has over a decade of experience in nanoscale science and technology. He is a specialist in dynamic AFM methods. He got his PhD from Leeds University in 2011 on the topic of AFM imaging of macromolecules on surfaces in air.

Lamiaa Elsherbiny

Lamiaa Elsherbiny is a PhD student in the Laboratory of Energy of Nanoscience (LENS) group in Khalifa University. She received her Bachelor in sustainable and renewable energy engineering at University of Sharjah. Her work at UoS consisted of modelling the optical properties of Dye Sensitized Solar Cells (DSSC). She received her master's degree in mechanical engineering at Khalifa University studying the forecasting performance of different machine learning algorithms of photovoltaic generation in Abu Dhabi. She is currently pursuing her PhD. degree in engineering by studying the impact of quantifying interaction forces on tuning the properties of surfaces using experimental and analytical methods.

Matteo Chiesa

Matteo Chiesa is the head of the LENS (Laboratory for Nano Science) at Khalifa University. He is also professor with the Arctic University of Norway UIT. His research focuses on the creation and implementation of technologies necessary to adapt the current energy system into a more sustainable, competitive and secure one, in particular the use of properly designed nano in solar energy systems. Prof. Chiesa's 15-years-long research efforts have been pivotal for Masdar in achieving photovoltaic projects promising electricity at record-low prices. With PPA (power purchase agreement) bids reaching as low as 1.35c/kWh, this surge of ultra-cheap installations marks a major milestone for PV technologies. In terms of scholarly contribution, Prof. Chiesa has built a dynamic team that is recognized by his community for consistently attempting the enhancement of the AFM (Atomic Force Microscopy) capability.

Symbols

m	Mass
z	Position of the tip relative to the cantilever
d	Tip-sample distance
t	Time
Q	Q factor
ω_0	Natural frequency
k	Spring constant
F_0	Magnitude of drive force
ϕ	Phase shift
F_{ts}	Tip-sample (surface) force
$\ddot{z}$	Acceleration
$\dot{z}$	Velocity
T	Period
W	Work
U	Potential energy
KE	Kinetic energy
E_T	Total energy
E_{avg}	Average energy
$\langle E \rangle$	Average energy
δ	Deformation
P_{dis}	Power dissipation
A	Amplitude
A_0	Free amplitude
W_D	Work done by the drive
V	Virial
E_{dis}	Energy dissipation
ω_r	Resonance frequency
ω_{eff}	Effective resonance frequency
ω_0'	Effective resonance frequency

Part I

Linear theory and applications in AFM

Chapter 1

Introduction

1.1 Oscillations, linear theory, and applications in AFM

The theory of oscillations can be studied from a mathematical point of view in terms of differential equations. The differential equation is written and then the solution or solutions worked out and mathematically analysed. Provided physical, economic, social, or other phenomena can be modelled in terms of equivalent differential equations, the solutions and results are applicable to all phenomena all the same. On the other hand, it is sometimes easier to learn a topic by having an experimental topic in mind. It is otherwise maybe surprising that a large body of phenomena in many fields of application will be easily understood if the equations are understood for a given case. The experimental analysis that forms the basis of this book is cantilever dynamics in dynamic atomic force microscopy (AFM). In a nutshell, the motion of the cantilever in dynamic AFM can be approximated to a perturbed driven oscillator. The generality of the analysis presented here can be confirmed by noting that much of what is covered in this book, particularly when dealing with the linear equation in part I, is similar to what is covered in generic expositions such as that by Tipler and Mosca [1] or Feynman's lectures on physics [2]. Maybe the main advantage of this exposition is that the linear and nonlinear theories of oscillators, particularly phenomena that can be reduced to the analysis of a point-mass on a spring, are discussed in detail and differences in terminology that could lead to doubt, clarified. This means that this book can be used as a textbook to teach oscillation theory with a focus on applications. This is possible because oscillations are present generally in physics, engineering, biology, economics, sociology, and so on. In summary, all phenomena dealing with oscillations can be reduced, to a first approximation, to a restoring parameter, i.e., force in mechanics, following Hooke's law. Since the AFM field is a niche in science, the general theory of oscillation does not cover the particularities of the field. This book covers such particularities. Thus, the book can be used as an introduction to dynamic AFM and oscillation theory that covers the terminology required to understand dynamic AFM. On the other

doi:10.1088/978-0-7503-5809-5ch1

hand, the ubiquity and generality of oscillation theory ensures that the book can also be used as a general introduction to oscillation theory at an undergraduate level. There is also advanced material, particularly in the second section of the book where nonlinearities are covered. For the sake of translationality, standard terminology is employed throughout when possible, especially in the first section. The first part, or section, (chapters 1–4) is based on the standard linear differential equations. The second part, or section, (chapters 5–8) is based on non-linear theory and covers advances in dynamic AFM over the past three decades (1990s–2022). Both sections are complimentary and employ standard terminology in oscillation theory when possible.

1.2 Oscillations in general

The oscillation of systems is a general phenomenon. A system oscillates if there is motion of the system around a neutral position. In dynamic AFM the physical system consists of a microcantilever which is made to oscillate around its position of equilibrium near a surface. A very sharp tip at the end of the cantilever periodically interacts with the surface via nanoscale forces. If the AFM cantilever vibrates far from the surface, i.e., $\sim 10–10^3$ nm away from the surface, in a vacuum, in air or in liquids, and provided the amplitude of oscillation is sufficiently small, i.e., $\sim 1–10^2$ nm, the oscillation is approximately linear, and the linear theory of oscillations applies. When the tip interacts closer to the surface, a nonlinear term, i.e., the tip-sample force F_{ts}, is introduced and the nonlinear theory applies. Typical systems that oscillate are a pendulum or waves deep in the ocean, i.e. water moves up and down around a neutral position. Importantly, the physics of vibrations, or oscillations, of mechanical systems can be extended to the oscillations of dynamic systems in fields ranging from atomic and molecular physics [3–5], biological systems [6–9], automobile mechanics [10], and economics [11]. Even the human body has a multitude of sensory organs responding to vibrations [8, 9]. In this respect, and while this book is concerned with oscillation theory based on the mass-point-spring model applied to AFM theory and experiment, the results are valid for any such system that can be reasonably well modelled as a point mass on a spring. The main mathematical trick consists of finding effective values for the parameters of the mathematical model. Provided such effective values can be found and provided the system can be reasonably well modelled as a mass-point-spring, the results are mathematically equivalent to the dynamics of oscillation of any general system. If these prerequisites are met, the oscillations of small, large, or even complex systems can be understood from the expressions, theory, and experiments discussed here. The first chapters deal with the linear theory, but this theory is shown to be crucial when considering nonlinear behaviour. It will be shown in part II, or section 2, (chapters 5–8) that the linear theory plays a very important role in the development of the mathematical formalism and interpretation of nonlinear behaviour. The nonlinear discussion treats oscillations affected by any nonlinear interaction, that is, by any general interaction including those emerging from conservative and dissipative phenomena. Finally, the transient response will not be discussed here in detail since the focus will be the steady state response, i.e., the system oscillating in equilibrium.

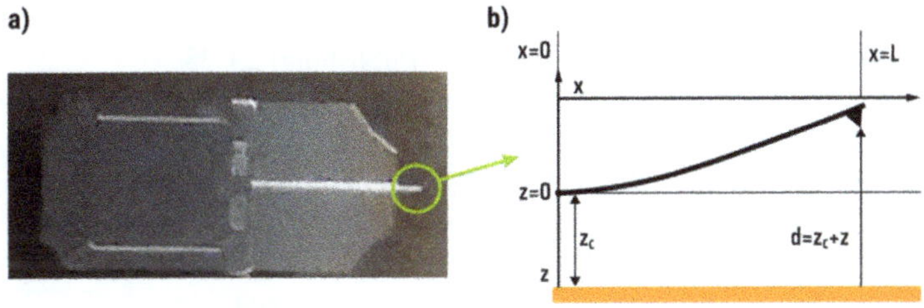

Figure 1.1. (a) Photo of a real AFM microcantilever and the cantilever holder. The microcantilever is some 100 μm in length (circle). (b) Simplified schematic of the microcantilever that allows one to model the system by exploiting beam theory [12–14].

An AFM microcantilever is shown in figure 1.1(a). In figure 1.1(b) we find a simplified schematic of the microcantilever on the basis of which a rheological model can be derived. A similar schematic is provided in figure 1.1 by Lozano and Garcia in their 2009 paper. This schematic (figure 1.1(b)) aims to provide an accurate description of the dynamic behaviour of the microcantilever. It is important when dealing with phenomena to draw such schematics since these allow deriving the main parameters in the model by visual inspection and by using simple algebra. Constraints can thus be found. From figure 1.1(b) the following geometric [12] parameters can be defined

z: tip position in the vertical axis,
z_c: cantilever equilibrium distance,
d: tip-sample distance,
x: cantilever position in the x axis,
L: length of the cantilever

References

[1] Tipler P A and Mosca G P 2003 *Physics for Scientists and Engineers* (New York: W.H. Freeman)

[2] Sands M, Feynman R and Leighton R 2011 *The Feynman Lectures on Physics* (New York : Basic Books)

[3] MacRae A U and Germer L H 1962 Thermal vibrations of surface atoms *Phys. Rev. Lett.* **8** 489–90

[4] Mayants L S and Shaltuper G B 1975 General methods of analysing molecular vibrations *J. Mol. Struct.* **24** 409–31

[5] Kučera O, Havelka D and Cifra M 2017 Vibrations of microtubules: physics that has not met biology yet *Wave Motion* **72** 13–22

[6] Parker K J, Huang S R, Musulin R A and Lerner R M 1990 Tissue response to mechanical vibrations for 'sonoelasticity imaging' *Ultrasound Med. Biol.* **16** 241–6

[7] Marvi M and Ghadiri M 2020 Retracted article: a mathematical model for vibration behavior analysis of DNA and using a resonant frequency of DNA for genome engineering *Sci. Rep.* **10** 3439

[8] Kasas S *et al* 2015 Detecting nanoscale vibrations as signature of life *Proc. Natl Acad. Sci.* **112** 378–81

[9] Zee E A 2020 The biology of vibration *Manual of Vibration Exercise and Vibration Therapy* ed J Rittweger (Cham: Springer International Publishing) 23–38
[10] Mansfield N J, Griffin J and M 2000 Difference thresholds for automobile seat vibration *Appl. Ergon.* **31** 255–61
[11] Blad M C and Christopher Zeeman E 1982 Oscillations between repressed inflation and keynesian equilibria due to inertia in decision making *J. Econ. Theory* **28** 165–82
[12] Santos S, Gadelrab K R, Souier T, Stefancich M and Chiesa M 2012 Quantifying dissipative contributions in nanoscale interactions *Nanoscale* **4** 792–800
[13] Graham K S 1993 *Fundamentals of Mechanical Vibrations* (New York: McGraw Hill)
[14] Steidel R 1989 *An Introduction to Mechanical Vibrations* 3rd edn (New York: Wiley)

IOP Publishing

Oscillations
Theory and applications in AFM

Tuza Adeyemi Olukan, Sergio Santos, Lamiaa Sami Elsherbiny and Matteo Chiesa

Chapter 2

Model description

2.1 Continuous and point-mass model descriptions of the cantilever-tip system

The motion of a system like that shown in figure 2.1 must be simplified and described in terms of a model that is both relatively easy to understand from a physical point of view and later allows quantifying the relevant parameters in a given experiment. As it turns out, this step is crucial to investigate oscillations and it is many times possible to reduce the motion of many dynamic systems to the spring-mass model (figure 2.1). Several groups in the AFM community [1–3] have shown that the well-known Euler–Bernoulli beam theory [4, 5] leads to the standard equation of the driven oscillator. Our group employed such theory to understand the implications of directly and indirectly driving the cantilever. In these papers the Euler–Bernoulli beam equation that governs the dynamics of a rectangular beam, equation (2.1) [6, 7], is described in terms of the relevant experimental AFM parameters. The reader can refer to any of the above papers to understand how the motion of the AFM cantilever, in particular the tip's position at $x = L$, can be modelled as the standard spring-mass model. For example, equations (27) and (33) by Lozano and Garcia [1] treat the Euler–Bernoulli beam equation, their equation (18), as follows:

$$EI\frac{\partial}{\partial x^4}\left[w(x,\,t) + a_1\frac{\partial w(x,\,t)}{\partial t}\right] + \rho bh\frac{\partial^2 w(x,\,t)}{\partial t^2} = -a_0\frac{\partial w(x,\,t)}{\partial t} + \delta(x - L)[F_{\text{exc}}(t) + F_{ts}(d)] \tag{2.1}$$

where E is the cantilever's Young's modulus, I is the area moment of inertia, a_1 is the internal damping coefficient, ρ is the mass density, b, h, and L are, respectively, the width, height, and length of the cantilever, a_0 is the hydrodynamic damping, $w(x,\,t)$ is the time-dependent vertical displacement of the differential beam's element

doi:10.1088/978-0-7503-5809-5ch2

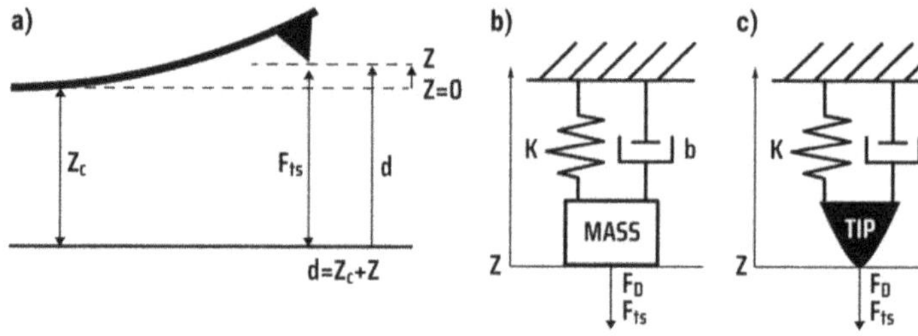

Figure 2.1. (a) Schematic of an AFM cantilever from which geometrical constraints can be derived. (b) and (c) Rheological models of the tip' motion. Both models are mathematically equivalent but the illustration in (c) showcases that the motion is discussed in terms of the tip.

placed at the x position, $F_{\text{exc}}(t)$ is the excitation force, and F_{ts} is the tip-sample force (compare with figure 2.1(b)).

The expression in (2.1) can be reduced to the standard differential equation describing the motion of a driven oscillator modelled as a mass and a spring (2.2)–(2.4). This example illustrates why understanding the standard linear theory is critical in order to understand AFM cantilever dynamics. There are many textbooks where the linear theory is described at the undergraduate level; for example, Tipler and Mosca's chapter on oscillations [8] or Feynman's lectures on physics, vol 1, chapters 21–23. The first part of this book is dedicated to the linear theory, and it is partly based on the two texts above with a focus on experiments in AFM. The second part is dedicated to the nonlinear theory, but it is still general.

To show that equation (2.1) is equivalent to a set of anharmonic differential equations, standard vibration theory can be exploited. In particular, equation (2.1) can be reduced to the following set of expressions [1–3, 6, 7]:

$$\ddot{Y}_m(t) + \frac{\omega_m}{Q_m}\dot{Y}_m(t) + \omega_m^2 Y_m(t) = \frac{F_m(t)}{m_m},\ m(\text{subscript}) = 1, 2, \ldots \tag{2.2}$$

The notation in equation (2.2) is that of Lozano and García [1] (refer to their equations (27) and (33)). While Lozano and García have used n as a subscript, we have used m (subscript) to refer to the set of modes because n is reserved for harmonics and because m typically stands for mode, i.e., note that Lozano and Garcia write n where we write m. $Y_m(t)$ is the time-dependent function of each eigenmode. In equation (2.2) m is the effective mass, Q is the effective quality factor, ω is the natural frequency, and $F(t)$ is the external force.

Finally, it can be shown that the expression in equation (2.2) can be reduced to equation (2.3) at $x = L$ where

$$\ddot{z}_m(t) + \frac{\omega_m}{Q_m}\dot{z}_m(t) + \omega_m^2 z_m(t) = \frac{F_{\text{exc}}(t) + F_{ts}(d)}{m},\ m(\text{subscript}) = 1, 2, \ldots \tag{2.3}$$

and where $z(t)$ is the modal projection of the tip motion. Equation (2.3) is a set of equations where we obtain one equation for each eigenmode m. In the rest of the chapter on linear theory only $m = 1$ will be considered but most of the derivations are valid for the higher modes. The textbooks by Mosca and Tipler and Feynman analyse such expressions, albeit $F_{ts} = 0$ in their analysis since only linear forces are considered. In summary, this book is dedicated to the analysis of the expressions in equation (2.3) with a focus on AFM experiments. On the other hand, the theory and considerations are, for the most part, general for other systems.

Since equation (2.3) is equivalent to the standard driven oscillator model typically discussed in textbooks, the standard rheological model in figure 2.1 applies. This means that the description and the analysis of this model are equivalent to the analysis of AFM theory. Such a model can be derived from the geometry of the AFM set up as shown in figure 2.1(a). All the parameters in figure 2.1 have been already defined.

2.2 Simple harmonic motion

If only the first mode of oscillation is considered, the full equation of motion governing the dynamics of the cantilever at the tip-sample junction can be reduced to [9–15]

$$\begin{aligned} m\ddot{z} + b\dot{z} + kz &= F_D + F_{ts} \\ m\ddot{z} + b\dot{z} + kz &= F_{\text{exc}} + F_{ts} \\ m\ddot{z} + b\dot{z} + kz &= F_0 \cos \omega t + F_{ts} \end{aligned} \tag{2.4}$$

where the first term is the net force, i.e., mass times acceleration, from Newton's laws of motion. The second term stands for viscous dissipation, the third term is the restoring force of the spring, and the terms on the right are the drive force $F_D(t) \equiv F_0 \cos(\omega t)$ and the tip sample interaction F_{ts}. We have written the drive force in many ways in equation (2.4) because the reader must pay attention to the definition given in each text. In particular, sometimes F_D is defined [16] as F_0 or as the amplitude [10, 17] of the drive A_D. The terminology is equivalent to that used by Tipler and Mosca in their chapter on oscillation under the subtitle resonance with the only main difference being that the last term is missing. In Feynman's lectures on physics, in chapter 23—Feynman's equation is written as equation (23.6)—but Feynman writes c rather than b. In order to avoid confusions, we note that Lozano and Garcia write F_D as $F_{\text{exc}}(t)$ as shown in equations (2.1) and (2.3) . The rest of chapters devoted to the linear theory deal with the analysis of equation (2.4) term by term with the exception of F_{ts}.

The first thing to note is that the rheological schematic in figures 2.1(b) and (c) can be used to derive equation (2.4) by applying Newton's laws.

Next, equation (2.4) is reduced to two terms only, giving

$$m\ddot{z} = -kz \tag{2.5}$$

where all the terms in equation (2.5) have been defined. That the terms have been formally defined does not mean that m or k are known in advance in a given experiment. Rather it is under the process known as calibration that the values of m and k will have to be found [18, 19]. This is standard routine in AFM, but also a routine that should be followed whenever dealing with oscillations modelled via equation (2.4). It is the mathematical exploration of equation (2.4) that will provide the tools to physically calibrate these parameters, i.e., to find effective values for these terms. Here, effective simply means that the system will behave 'as if' it had mass m and spring constant k. That is, since m only represents the effective mass of the system, m does not refer to the actual mass of the cantilever. Using experimental data to determine the value of m is what ensures the model's consistency. The same can be said about the rest of the parameters involved in equation (2.4). The rheological model for equation (2.5) is shown in figure 2.2.

The exploration of equation (2.5) provides an insight into the physical meaning of the terms in the equation. Otherwise equation (2.5) is typically known as the equation describing simple harmonic motion. Considering equation (2.5) alone, the only force acting on the mass is the restoring force

$$\begin{aligned} F &= -kz \\ m\ddot{z} &= -kz \end{aligned} \tag{2.6}$$

Equation (2.6) is Newton's equation for representing the model in figure 2.2. The physical meaning of such force is that whenever the object of mass m is displaced from its equilibrium position $z = 0$, the force tends to bring it back to $z = 0$. For this reason, the force in equation (2.6) is typically termed 'restoring force', i.e., it tends to restore the mass to its equilibrium position. This is the typical behaviour of a spring. The spring might further act on extension or compression, but they always tend to restore the equilibrium position. Provided the restoring force is proportional to the displacement, the restoring force is linear as shown in equation (2.6). At this point it is worth clarifying that under this model (figures 2.1 and 2.2) the object is an ideal point of mass m. This ideal point can represent any physical system of finite size.

It is essential to understand how the 'natural frequency', ω_0, which is an angular frequency given in radians, is derived from this equation. The meaning of 'natural' will be discussed as the analysis proceeds.

The position of the point of mass m is determined as z and the behaviour of the mass is parametrized through equations (2.7)–(2.9).

$$z = A\cos(\omega t + \phi) \tag{2.7}$$

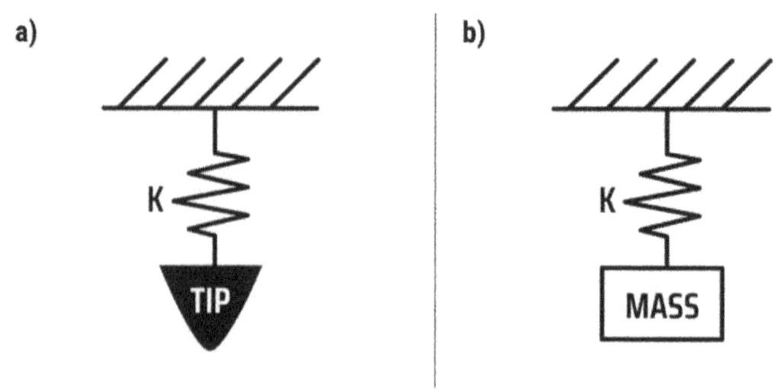

Figure 2.2. (a) and (b) Rheological models of the tip' motion. Both models are mathematically equivalent but the illustration in (a) showcases that the motion is discussed in terms of the tip.

$$\ddot{z} = -A\omega^2 \cos(\omega t + \phi) \tag{2.8}$$

$$\ddot{z} = -\frac{k}{m} z \tag{2.9}$$

where A is the amplitude of oscillation and where the phase (shift) ϕ depends on where $t = 0$ is defined.

In short, equations (2.7)–(2.9) represent the solution of equation (2.5) as confirmed by substitution. Inserting equations (2.7) and (2.8) into equation (2.5) gives

$$-mA\omega^2 \cos(\omega t + \phi) = -kA \cos(\omega t + \phi) \tag{2.10}$$

$$-m\omega^2 = -k \tag{2.11}$$

where the above are solutions provided the identity below is true

$$\omega^2 \equiv \frac{k}{m} \tag{2.12}$$

Equation (2.12) is the first important result since k and m can now be related to a parameter that can be experimentally measured, namely, the angular frequency at which the system will oscillate. The above identity is employed to define the natural frequency of oscillation.

$$\omega^2 = \frac{k}{m} \equiv \omega_0^2 \tag{2.13}$$

The frequency f in Hertz is given as usual

$$f_0 = \frac{\omega_0}{2\pi} \tag{2.14}$$

The spring constant is thus related to the mass and the angular frequency as follows:

$$\omega_0 = \sqrt{k/m} \tag{2.15}$$

Equation (2.15) is the expression obtained when there is no dissipative term. This indicates that when the system resonates under the influence of no other force, for instance, in a high vacuum, a mass would move up and down sinusoidally at its natural frequency as given by equations (2.12)–(2.15). The definition of the natural frequency further shows that it is given by conservative forces alone. This is because $-kz$ is a conservative force.

The main result of the analysis in this section is that

1. When the only force on the mass is the restoring force, the tendency is to oscillate sinusoidally at frequency f_0 or ω_0.
2. The frequency of oscillation is termed 'natural frequency' and it is determined by the effective mass m and the effective spring constant k.
3. Importantly, if there is nothing that disturbs the motion of the system, the mass will always remain in equilibrium.

When the mass moves outside equilibrium $z = 0$ there is sinusoidal motion as described by equation (2.7). The description of such motion is left for the next chapters. It is also possible to understand the concepts of potential energy and kinetic energy from equation (2.5) alone, but such analysis is also left for the next chapters.

2.3 Damped oscillations

In this section, there are several objectives.

1. To extend the analysis to the situation where there is damping or dissipation. In particular we will deal with standard viscous damping. Viscous damping can be understood as friction within a medium, i.e., it opposes the motion and therefore has a tendency to stop the body [20].
2. To investigate the effects of dissipation. The addition of viscosity has two important effects. First it displaces the point of maximum amplitude. This implies that the system will now resonate at $\omega_r \neq \omega_0$, i.e., there is maximum amplitude at a frequency different to ω_0. Second, the system will respond with significant values of amplitude A at a range of frequencies around ω_r.

In short, the aim is to relate the frequency of resonance ω_r to the natural frequency ω_0 when there is viscous dissipation, such as air or liquids. In AFM, viscosity is present in air environments due to friction with air while in liquid environments there is friction due to the liquid. The natural frequency is related to the natural period T as shown in equation (2.16).

$$f_0 = \frac{1}{T_0} = \frac{1}{\frac{2\pi}{\omega_0}} = \frac{\omega_0}{2\pi} \equiv \frac{1}{2\pi}\sqrt{\frac{k}{m}} \tag{2.16}$$

To find a relationship between the resonance frequency and the natural frequency we begin with equation (2.17). This equation is derived from the rheological model shown in figure 2.3. This simplified model is also discussed in detail in standard textbooks, for example by Tipler and Mosca and in Feynman's lectures.

$$m\ddot{z} + b\dot{z} + kz = 0 \tag{2.17}$$

First, the term b is sometimes written in terms of Q where Q is defined as

$$\gamma = \frac{\omega_0}{Q} \tag{2.18}$$

and

$$b = m\frac{\omega_0}{Q} \equiv m\gamma \tag{2.19}$$

It is the task of this chapter to go through the derivation and physical interpretation of equations (2.18) and (2.19). In AFM the typical parameters that are calibrated during every experiment and routinely are the spring constant k, the resonant

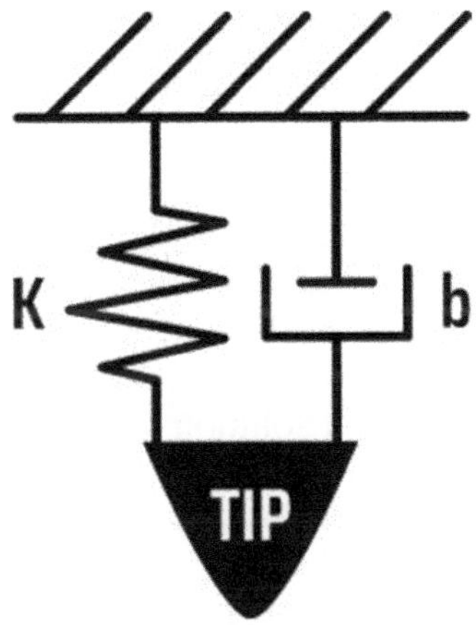

Figure 2.3. Rheological models of the tip's motion in the presence of viscosity.

frequency ω_r and Q. Through these expressions the equation of motion equation (2.17) can be written in the following equivalent forms

$$\frac{k}{\omega_0^2}\ddot{z} + \frac{k}{\omega_0 Q}\dot{z} + kz = 0 \tag{2.20}$$

$$m\frac{d^2z}{dt^2} + \frac{m\omega_0}{Q}\frac{dz}{dt} + kz = 0 \tag{2.21}$$

$$\ddot{z} + \frac{b}{m}\dot{z} + \omega_0^2 z = 0 \tag{2.22}$$

Equations (2.17) and (2.20)–(2.22) are mathematically equivalent and exploit the relationships in equations (2.17), (2.18), and (2.19). As the analysis of the linear system proceeds, further relationships are obtained, and these are incorporated into the theory. For this reason, it is important to go through the full analysis.

In order to find a solution to equation (2.22) the following solution is proposed:

$$z = Be^{i\omega t} \equiv x + iy \tag{2.23}$$

Since the equation of motion is linear, it follows that if $z = x + iy$ is a solution of equation (2.17) so are x and y.

Time derivatives produce the following result

$$z = Be^{i\omega t} \tag{2.24}$$

$$\dot{z} = i\omega Be^{i\omega t} \tag{2.25}$$

$$\ddot{z} = -\omega^2 Be^{i\omega t} \tag{2.26}$$

Substituting into equation (2.17),

$$-\omega^2 Be^{i\omega t} + \frac{b}{m}i\omega Be^{i\omega t} + \omega_0^2 Be^{i\omega t} = 0 \tag{2.27}$$

Dividing by $Be^{i\omega t}$ results in the following expression

$$-\omega^2 - \frac{b}{m}\mathrm{i}\omega + \omega_0^2 = 0 \tag{2.28}$$

Now we have an equation that is quadratic in terms of ω. The equation is telling us that the system will resonate at $\omega \equiv \omega_r$ where the maximum value of A will be found at ω_r. This is a consequence of the presence of viscous damping, i.e., of the second term in equation (2.17). Since equation (2.28) contains ω_0, this solution provides a method to relate ω_r to ω_0 provided a solution ω is found.

Solving equation (2.28),

$$\omega^2 + \frac{ib\omega}{m} - \omega_0^2 = 0$$

$$\omega = \frac{ib}{2m} \pm \frac{1}{2}\sqrt{\left(\frac{ib}{m}\right)^2 + 4\omega_0^2}$$

$$\omega = \frac{b}{2m}i \pm \omega_0\sqrt{1 - \left(\frac{b}{2m\omega_0}\right)^2} \tag{2.29}$$

From linearity it follows that equation (2.29) contains all the solutions to equation (2.17) in

$$\omega = x + iy$$

This means that both x and y are solutions. Thus, the following is a solution ($\omega_r \equiv y$) to equation (2.17)

$$\omega_r = \omega_0\sqrt{1 - \left(\frac{b}{2m\omega_0}\right)^2} = \omega_0\sqrt{1 - \frac{1}{4Q^2}} \tag{2.30}$$

Equation (2.30) explains that the resonance frequency will decrease with increasing dissipation b or decreasing Q. Only in the limit $Q \to \infty$, $\omega_r = \omega_0$. This is equivalent to saying $b = 0$.

In AFM, frequency sweeps are taken to determine ω_r. Such frequency sweeps are taken high above the surface, i.e., $z_c \sim \mu\mathrm{m}$. This means that taking a frequency sweep near the surface or far from the surface will result in different values of ω_r if dissipation, i.e., b or Q, varies as a function of proximity to the surface. In other words, the result in equation (2.30) allows determining viscosity as a function of cantilever-surface proximity based on the shift from ω_0. The closer ω_r is to ω_0, the lower the viscosity. Viscosity commonly increases as one approaches the surface. Consequently, one should observe a lower resonance frequency, that is, the peak in amplitude A should be found at lower frequencies, the closer the cantilever is brought to the surface. A standard frequency sweep performed with a cypher AFM is shown in figure 2.4.

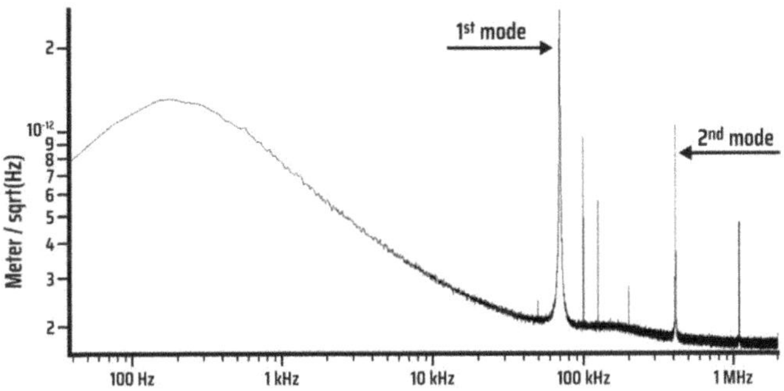

Figure 2.4. Standard frequency sweep for an AFM cantilever.

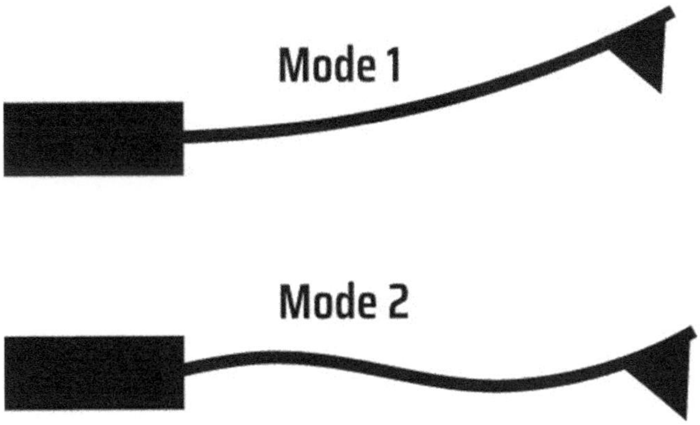

Figure 2.5. Illustration of the first two modes of a cantilever

In such a sweep the amplitude A is shown for each value of ω from $0 < f < 400$ kHz (approx.). The idea is to get the value ω_r where ω_r is defined when the amplitude reaches its maximum value. For values of Q that are relatively large, i.e., ~10, 100, $\omega_r \approx \omega_0$ as observed by inspecting equation (2.30). For example, for $Q = 100$ and $f_0 = \omega_0/2\pi = 100$ kHz, $f_r = 99.999$ kHz which is a 0.1% shift in frequency.

In ultra-high vacuum, typical values of Q are ~10 000. In air environments $Q \sim 100$ and in liquid environments we experimentally find that $Q \sim$ 1–10. In figure 2.5 there are several main peaks. The first peak corresponds to the resonance of the first mode while the second corresponds to the resonance of the second mode. The peak of the third mode is also observed near 1 MHz. The other peaks in figure 2.4 are due to noise. This can be experimentally confirmed by exploring the graph with AFM software, but it has not been shown here. Physically, the cantilever will move very differently when resonating near the first and second modes. In figure 2.5 the typical behaviour of modes 1 and 2 is illustrated.

References

[1] Lozano J R and Garcia R 2009 Theory of phase spectroscopy in bimodal atomic force microscopy *Phys. Rev.* B **79** 014110

[2] Butt H-J, Cappella B and Kappl M 2005 Force measurements with the atomic force microscope: technique, interpretation and applications *Surf. Sci. Rep.* **59** 1–152
[3] Butt H-J and Jaschke M 1995 Calculation of thermal noise in atomic I force microscopy *Nanotechnology* **6** 1–7
[4] Graham K S 1993 *Fundamentals of Mechanical Vibrations* (New York: McGraw Hill)
[5] Steidel R 1989 *An Introduction to Mechanical Vibrations* 3rd edn (New York: Wiley)
[6] Gadelrab K, Santos S, Font J and Chiesa M 2013 Single cycle and transient force measurements in dynamic atomic force microscopy *Nanoscale* **5** 10776–93
[7] Santos S, Gadelrab K, Font J and Chiesa M 2013 Single-cycle atomic force microscope force reconstruction: resolving time-dependent interactions *New J. Phys.* **15** 083034
[8] Tipler P A and Mosca G P 2003 *Physics for Scientists and Engineers* (New York: W.H. Freeman)
[9] Giessibl F J 1997 Forces and frequency shifts in atomic-resolution dynamic-force microscopy *Phys. Rev.* **B 56** 16010
[10] Anczykowski B, Gotsmann B, Fuchs H, Cleveland J P and Elings V B 1999 How to measure energy dissipation in dynamic mode atomic force microscopy *Appl. Surf. Sci.* **140** 376–82
[11] Garcia R and San Paulo A 1999 Attractive and repulsive tip-sample interaction regimes in tapping-mode atomic force microscopy *Phys. Rev.* **B 60** 4961
[12] Santos S, Barcons V, Verdaguer A and Chiesa M 2011 Subharmonic excitation in amplitude modulation atomic force microscopy in the presence of adsorbed water layers *J. Appl. Phys.* **110** 114902
[13] Amadei C A, Tang T C, Chiesa M and Santos S 2013 The aging of a surface and the evolution of conservative and dissipative nanoscale interactions *J. Chem. Phys.* **139** 084708
[14] Xu X and Raman A 2007 Comparative dynamics of magnetically, acoustically, and brownian motion driven microcantilevers in liquids *J. Appl. Phys.* **102** 034303–8
[15] Melcher J *et al* 2009 Origins of phase contrast in the atomic force microscope in liquids *Proc. Natl Acad. Sci.* **106** 13655–60
[16] Hu S and Raman A 2008 Inverting amplitude and phase to reconstruct tip–sample interaction forces in tapping mode atomic force microscopy *Nanotechnology* **19** 375704
[17] Cleveland J P, Anczykowski B, Schmid A E and Elings V B 1998 Energy dissipation in tapping-mode atomic force microscopy *Appl. Phys. Lett.* **72** 2613–5
[18] Sader J E *et al* 2012 Spring constant calibration of atomic force microscope cantilevers of arbitrary shape *Rev. Sci. Instrum.* **83** 103705–16
[19] Proksch R, Schäffer T, Cleveland J, Callahan R and Viani M 2004 Finite optical spot size and position corrections in thermal spring constant calibration *Nanotechnology* **15** 1344
[20] Morini G L 2008 'Viscous dissipation,' *Encyclopedia of Microfluidics and Nanofluidics* ed D Li (Boston, MA: Springer) pp 2155–64

IOP Publishing

Oscillations
Theory and applications in AFM
Tuza Adeyemi Olukan, Sergio Santos, Lamiaa Sami Elsherbiny and Matteo Chiesa

Chapter 3

Energy

3.1 Relationship between stored energy and kinetic energy

This section covers the relationship between total stored energy E, potential energy U, and kinetic energy KE through the analysis of the expression of simple harmonic motion. It is typical in textbooks to describe the relationship between stored energy and kinetic energy right after discussion of equation (2.5), that is, after the equation describing simple harmonic motion. See for example Tipler and Mosca's discussion on oscillations or the discussion provided by Feynman in his lectures. This is because if there is dissipation and no energy enters the system, for example by driving it with an external force F_d, the amplitude A will eventually reach 0, i.e., the system will stop. The rheological model describing equation (2.5) is shown again as figure 3.1 with the addition of an oscillatory sketch of amplitude A, i.e., $z = A\cos(\omega_0 t + \phi)$. The period of oscillation T is simply

$$T_0 = \frac{1}{f_0} \tag{3.1}$$

If there is no drive, in order to reach the amplitude A the spring must be extended to $z = A$. Then, it will oscillate with amplitude A. Since there is no dissipation, the energy stored at $z = A$ and $\dot{z} = 0$ will remain in the system with time, i.e., $\frac{dE}{dt} = 0$.

The total energy of the system is the sum of the potential energy and the kinetic energy. Using Hook's Law, the instantaneous potential energy stored in the spring as a function of time can be found as follows

$$F = -kz$$

$$dW = F \cdot dz \equiv du$$

$$W \equiv U = \int_{z_1}^{z_2} kzdz$$

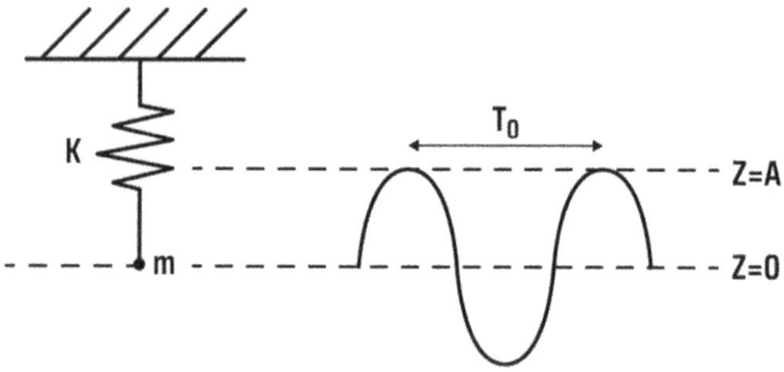

Figure 3.1. A rheological model describing equation (2.5) is shown again together with an oscillatory sketch of amplitude A and period T_0.

$$W \equiv \Delta U = \frac{1}{2}kz_2^2 - \frac{1}{2}kz_1^2$$

$$U = \frac{1}{2}kz^2 + C \tag{3.2}$$

where dW is the differential of work done by the restoring force per differential displacement dz, the total work done from $z = z_1$ to $z = z_2$ is W which can be identified as the difference in potential energy ΔU from $z = z_1$ to $z = z_2$. The stored energy in the spring U is therefore arbitrarily defined. If the displacement occurs from $z = 0$ to $z = z \equiv A$, then, as measured from $z = 0(U = 0)$,

$$U_0 = \frac{1}{2}kA^2 \tag{3.3}$$

where the zero potential energy is also arbitrarily defined at $z = 0$. For any arbitrary defined potential energy

$$U = \frac{1}{2}kz^2 + C \tag{3.4}$$

where the zero potential energy is arbitrarily as C at $z = 0$. Typically, $C = 0$. The arbitrariness of defining $U = C = 0$ at $z = 0$ is due to the fact that only differences in potential energy are physically meaningful and it is useful to define potential energy in terms of the range of motion, i.e., range in z is $2A$ where $-A \leqslant z \leqslant A$. Furthermore, that the potential energy is zero when $-kz = 0$ is also useful.

Substituting equation (2.7) into equation (3.4):

$$z = A\cos(\omega t + \phi)$$

$$U = \frac{1}{2}kA^2\cos^2(\omega t + \phi) \tag{3.5}$$

Equation (3.5) defines the instantaneous potential energy as a function of position z and $z = z(t)$. The kinetic energy KE can be expressed as follows:

$$\dot{z} = -\omega A\sin(\omega t + \phi)$$

$$KE = \frac{1}{2}mv^2 = \frac{1}{2}m\dot{z}^2$$

$$KE = \frac{1}{2}mA^2\omega^2 \sin^2 (\omega t + \phi) \tag{3.6}$$

Using equation (2.12) and provided $\omega = \omega_0$

$$KE = \frac{1}{2}kA^2 \sin^2 (\omega t + \phi) \tag{3.7}$$

The total energy of the system E_T is the sum of the potential and kinetic energy. Then, combining equation (3.5) and (3.7)

$$E_T = KE + U$$

$$E_T = \frac{1}{2}kA^2(\cos^2(\omega t + \phi) + \sin^2 (\omega t + \phi))$$

$$E_T = \frac{1}{2}kA^2 \tag{3.8}$$

where A is the amplitude of oscillation and k is the spring constant. The average kinetic energy of the system can be computed by integrating equation (3.7).

$$KE_{\text{avg}} = \frac{1}{T}\int_0^t \frac{1}{2}kA^2 \sin^2 (\omega t + \phi)dt \tag{3.9}$$

The following are useful relationships to solve equation (3.9):

$$\sin(A + B) = \sin A \cos B + \cos A \sin B$$

$$\sin (\omega t + \phi) = \sin \omega t \cos \phi + \cos \omega t \sin \phi$$

$$\sin^2 (\omega t + \phi) = \sin^2 (\omega t) \cos^2 \phi + \sin \omega t \cos \omega t \sin \phi \cos \phi + \cos^2 (\omega t) \sin^2 \phi$$

$$+ \sin \omega t \cos \omega t \sin \phi \cos \phi$$

These integrals have solutions that follow from orthogonality. See for example the chapters on Fourier series by KA Stroud [1]. The average kinetic energy is equal to the sum of the integrals I_1 to I_4

$$kE_{\text{avg}} \equiv \langle KE\rangle = \frac{1}{T}\frac{1}{2}KA^2(I_1 + I_2 + I_3 + I_4)$$

where

$$I_1 = \cos^2 \phi \int_0^T \sin^2 (\omega t)dt = \frac{T}{2} \cos^2 \phi$$

$$I_2 = \sin^2 \phi \int_0^T \cos^2 (\omega t)dt = \frac{T}{2} \sin^2 \phi$$

$$I_3 = I_4 = \frac{1}{2}\left[\int_0^T \sin\omega t \cos\omega t \; dt\right]\sin\phi\cos\phi$$

$$I_3 = I_4 = 0$$

$$\langle KE\rangle = \frac{1}{T}\frac{1}{2}KA^2\frac{T}{2}[\sin^2\phi + \cos^2\phi]$$

$$\sin^2\phi + \cos^2\phi = 1$$

$$\langle KE\rangle = \frac{1}{2}\left[\frac{1}{2}KA^2\right] = \frac{1}{2}E_T \tag{3.10}$$

Moreover, the average potential energy can be expressed as follows:

$$U_{\text{avg}} \equiv \langle U\rangle = \frac{1}{2}kA^2\frac{1}{T}\int_0^T \cos^2(\omega t + \phi)dt = \frac{1}{2}E_T \tag{3.11}$$

The above integrals are solved in a similar way to that described to get to equation (3.10). To understand the relationship between the average kinetic and potential energies, one can assume that the velocity is zero and see what happens to KE and U. For $\dot{z} = 0$

$$KE = \frac{1}{2}m(\dot{z})^2 = 0 \tag{3.12}$$

Equation (3.12) implies that the total energy when $\dot{z} = 0$ is found as U. This will yield maximum potential energy at maximum displacement $z = A$ when $\dot{z} = 0$

$$U\,|_{\substack{z=A\\ \dot{z}=0}} = \frac{1}{2}kz^2 = \frac{1}{2}kA^2 \tag{3.13}$$

Assuming the position is zero, the potential energy U is zero and all the energy is found a kinetic energy. Since KE is proportional to the velocity squared, the velocity is maximum at $z = 0$

$$U\,|_{\substack{z=0\\ \dot{z}=\max}} = \frac{1}{2}k(z)^2 = 0 \tag{3.14}$$

$$KE = \frac{1}{2}m\dot{z}^2 = \frac{1}{2}kA^2 \tag{3.15}$$

From equation (2.15), it follows that at $z = 0$ the maximum velocity is found, and it can be written as

$$\dot{z}(\max) = \omega_0 A \tag{3.16}$$

This result is identical to what directly follows from the expression from instantaneous velocity

$$\dot{z} = -A\omega_0 \sin(\omega t + \phi) \tag{3.17}$$

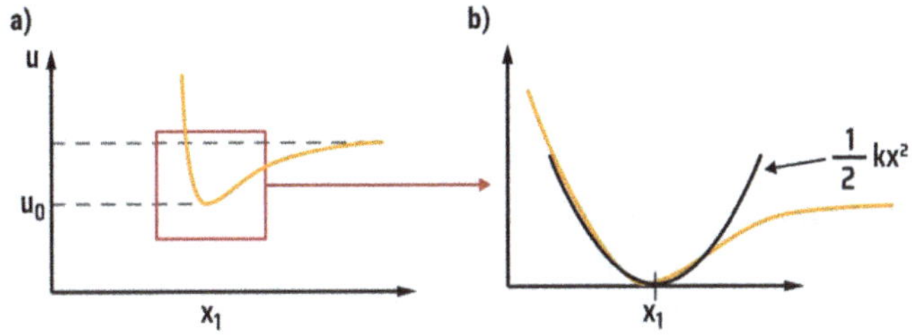

Figure 3.2. (a) Schematic illustrating the relationship between the potential energy U between two hydrogen atoms and its equilibrium position. (b) Schematic showing U as a function of x for a perfect Hookean potential, i.e., $\frac{1}{2}kx^2$, and the true potential between the two atoms. As observed the two potentials are similar when $x \approx x_1$.

The above discussion is typical of the analyses on energy in simple harmonic motion. To exemplify the generality of this discussion, figure 3.2 shows the potential energy U function versus the separation of two hydrogen atoms. The assumption is that the force near equilibrium for the two atoms is approximately Hookean (3.2). A similar example is discussed by both Tipler and Mosca and by Feynman when discussing oscillations. This is not coincidental, rather, it shows that is it is useful to describe atomic phenomena in terms of simple atomic motion and in particular to interpret the phenomena in terms of energy.

The expression used to describe the force between atoms as shown in figure 3.2 can be found by differentiating the function U (3.4) with distance x (or equivalently z). Then

$$F_x = -\frac{dU}{dx} = -k(x - x_1) \tag{3.18}$$

where $x = x_1$ is defined as the neutral position for the two atoms, i.e., $-kx = 0$. The implication is that the separation between the atoms will oscillate sinusoidally around x_1. In figure 3.2(a) the 'true' potential energy for the atoms is shown as a function of separation x. For the case of hydrogen atoms discussed by Tipler and Mosca $x_1 = 0.74$ nm. As shown in the scheme in figure 3.2(b), the quadratic expression for U (3.4) is only an approximation of the true potential energy between the atoms around x_1.

3.2 Distance independent or constant forces

This section is dedicated to the addition of a constant force to the system. We recall that the investigation of equation (2.4) is the main focus of this text. This equation is written here again as equation (3.19) to discuss the terms already analyzed in Chapter 2. Further analysis of the equation will be conducted in this chapter.

$$m\ddot{z} + \frac{m\omega_0}{Q}\dot{z} + kz = F_0 \cos \omega t + F_{ts} \tag{3.19}$$

In summary, equation (3.19) is the full governing equation of motion of the cantilever that also allows us to understand the tip-sample force F_{ts} since the force is also represented. The main assumption so far is that the analysis is carried out for the first mode $m = 1$.

So far the third term, i.e., kz, has been analysed in sections 2.2 and 2.3. The dissipative term $\frac{m\omega_0}{Q\dot{z}}$ has been discussed in section 2.3. The terms on the right side of the equation have not been discussed yet. Only the drive force is discussed in the next chapter. The term F_{ts} is added in the second section. Before proceeding to those terms the following equation of motion will be considered

$$m\ddot{z} + kz + F_C = 0 \tag{3.20}$$

Tipler and Mosca discuss a similar example where the constant force F_c is represented by the gravitational force, i.e. $F_c = mg$. The force mg is distance independent in that both g and m are independent of position z. This is of course true for variations in z in the order of meters or less since it is well known that g depends on the distance with respect to the centre of the Earth and these changes become significant when considering many km.

The model represented by the differential equation in equation (3.20) turns out to be meaningful in AFM since the tip-sample force in air might display a 'plateau' in the proximity of the surface. Figure 3.1 shows experimental force profiles [2]. F_{ts} as a function of tip-sample distance d but a plateau can be seen in one of the profiles. Here F_{ts} is normalised where minima are identified as −1. Minima in force in AFM are typically identified with the adhesion force F_{AD}. The force in black (figure 3.3) is a typical force profile where the force is negative, i.e., the tip and the sample attract each other, for $d > 0$. For smaller values of d the force increases with decreasing distance. The force shown in black was obtained on a cleaved mica surface. On the other hand the force in orange displays a 'plateau' of 2 nm where $F_{ts} \sim F_{AD}$. In this region the force is approximately constant. In 2013 we termed this region the SASS region [3] standing for small amplitude and small set point regime [2, 4–6].

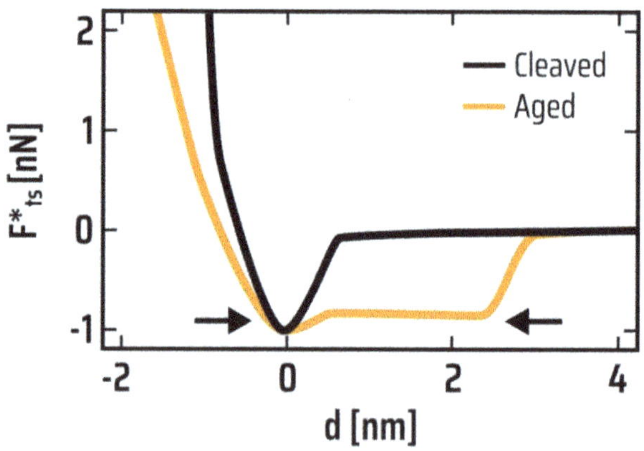

Figure 3.3. Force profiles where F_{ts} is plotted as a function of distance d. F_{ts} is normalized $F_{ts}^* = \frac{F_{TS}}{F_{AD}}$ by dividing by minima. The profile in black is for a tip interacting with a freshly cleaved mica surface and the orange one with an aged surface. Reproduced from [2] with permission from the Royal Society of Chemistry.

The expected or standard force profile in air environments should be similar to that of the cleaved surface in figure 3.3 (black lines) [7] but in air environments we find that the plateau shown in orange lines might form with time [7–11]. It has been hypothesized that this plateau is a consequence of the aging of the surface whereby a nanometric layer of water and other contaminants forms on surfaces [8, 10–12]. A schematic of this phenomenon is shown in figure 3.4(a). A schematic of how the tip of an AFM would oscillate in the region of constant force is shown in figure 3.4(b).

The rheological model for the constant force is shown in figure 3.5. The following forces must be considered

$$m\frac{d^2z}{dt^2} = -kz + F_C \tag{3.21}$$

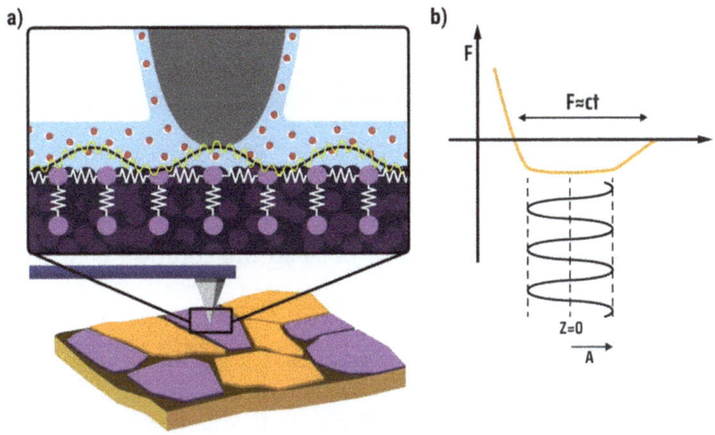

Figure 3.4. Illustration and schematic of a tip vibrating near a surface inside a hydration layer of water of height in the range of 1–2 nm. The force profile shows that the tip might oscillate within this layer where the force can be approximately constant.

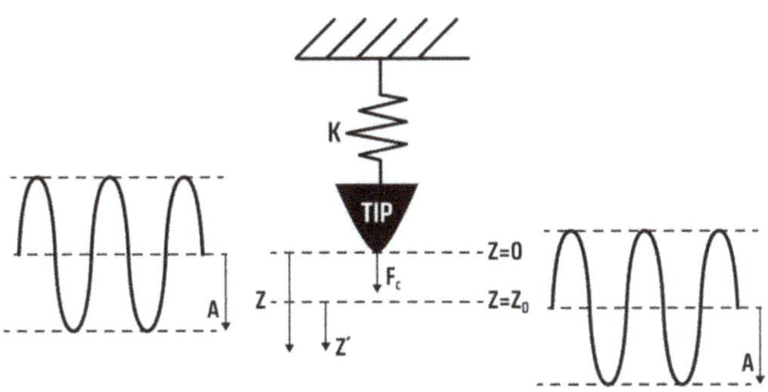

Figure 3.5. Schematic of the shift in reference from $z = 0$ to $z' = 0$ when $z = z_0$ due to the addition of a constant force F_C.

The value of F_c can be written in terms of k since in equilibrium and in the absence of oscillation

$$0 = -kz + F_C \quad (3.22)$$

$$kz = F_C \quad (3.23)$$

Equation (3.24) allows us to write a new equilibrium position and coordinate axis z',

$$z = \frac{F_C}{k} \equiv z_0(z_0 \Rightarrow z' = 0) \quad (3.24)$$

The above expression (3.24) can be used to determine the mean cantilever deflection z_0. This deflection is due to an average force $\langle F_{ts} \rangle$ equivalent to a constant force F_c in its effects on z_0. With the new equilibrium position at $z = z_0$ a new axis z' can be found from the diagram in figure 3.5. Then

$$m\frac{d^2(z' + z_0)}{dt^2} = -k(z' + z_0) + F_C = -kz' - kz_0 + F_C$$

$$m\frac{d^2z'}{dt^2} = -kz' \quad (3.25)$$

This expression is equivalent to equation (2.5) so the solution is also the same

$$z' = A\cos(\omega t + \phi) \quad (3.26)$$

The main result of adding a constant force is that if the tip oscillates in a region of constant force like the SASS region shown in figures 3.3 and 3.4, the behaviour is equivalent to simple harmonic motion. The only terms missing in this equation are the drive F_D, and dissipation.

The potential energy in this case is

$$U = U_{\text{spring}} + U_{\text{constant force}} = \frac{1}{2}kz'^2 \quad (3.27)$$

The maximum potential energy is still

$$U = \frac{1}{2}kA^2\left(U = 0 \text{ at } z' = 0 \text{ or } z = z_0\right) \quad (3.28)$$

In SASS imaging, and assuming the ideal condition where $F_{ts} =$ constant in the region, the force of adhesion can be found as [13]

$$F_{ts} \approx F_{AD} \approx kz_0 \quad (3.29)$$

In reality, since the amplitude of oscillation A might be larger than the range for which $F_{ts} =$ constant (figure 3.4), the above expression is only a linear approximation. A more realistic interpretation of the phenomenon in SASS can be understood by considering the illustration in figure 3.6 [6]. In the figure, an example is shown of the

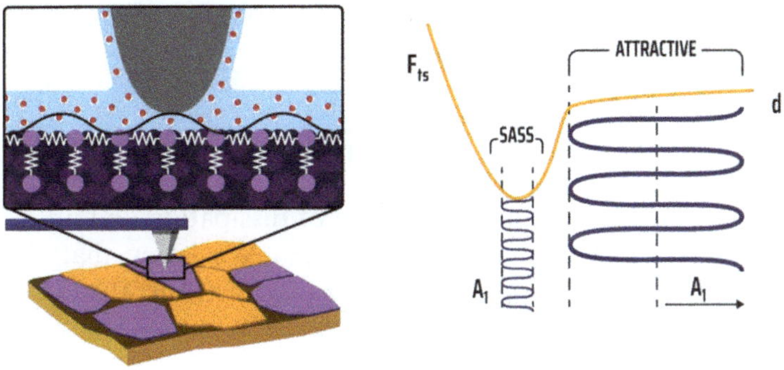

Figure 3.6. Schematic similar to that in figure 3.4 where an AFM tip oscillates inside a hydration layer and where relatively realistic force profiles are shown depicting how the tip can be made to oscillate in the attractive regime or within the well, i.e., in perpetual contact with the water layer.

phenomenon where the cantilever interacts with the sample in the attractive mode only, i.e., $F_{ts} < 0$, with relatively large amplitudes A. Here, it is clear that the force is not constant. Another example is shown in the figure where the SASS mode of imaging occurs where $F_{ts} \approx F_{AD}$ and the oscillation amplitude is sufficiently small. As shown, it is possible that even if the force is not constant, for example, at the edges as also shown in figure 3.6 representing SASS imaging, the tip oscillates mostly and approximately at the bottom of the well. We see that the conditions are that [3, 14]

(1) The range for which the $F_{ts} \approx$ constant must be approximately $2A$, i.e., twice the amplitude of oscillation.
(2) Second, the tip must be made to oscillate at the bottom of the well.

The first condition can be easily achieved since AFM allows the user to set any oscillation amplitude. The second condition is not trivial from an experimental point of view [14]. We have recently shown that these conditions are still to be fully elucidated [5, 15]. It might be worth exploring these conditions in the future since they might lead to high resolution imaging [16, 17] both because (1) of the small A conditions and because (2) the tip is in very close proximity to the sample where resolution is enhanced [18].

3.3 *Q* factor and dissipation

The Q factor has been defined in section 2.3 and its relationship to dissipation has been already shown to some detail. Here, some physical interpretations will be discussed in terms of the derivations resulting from the equation of motion.

The equation of motion that accounts for dissipation (equation (2.17)) is rewritten here

$$-kz - b\frac{dz}{dt} = m\frac{d^2z}{dt^2} \tag{3.30}$$

where all the terms have been already defined. The solution of this equation in terms of ω has been shown (equation (2.30)) to be

$$\omega_r = \omega_0\sqrt{1 - \frac{1}{4Q^2}} \tag{3.31}$$

This is the resonant frequency in the presence of dissipation and no drive. Clearly, for any finite value of Q we have that $\omega_r < \omega_0$. We know that the solution is sinusoidal where $z \propto \cos(\omega t + \phi)$. In order to find the general solution (equations (2.23) and (2.24)) were assumed to hold. The same assumptions are shown to aid the discussion

$$z = A_0 e^{i\omega t} \equiv z + iy \tag{3.32}$$

$$z = A_0 e^{i\omega t} \tag{3.33}$$

where the general solution can be written as before as

$$\omega = \frac{b}{2m} i \pm \omega_r \tag{3.34}$$

Inserting (3.34) into (3.33)

$$z = A_0 e^{i\left(\frac{b}{2m}i + \omega_r\right)} = A_0 e^{-\frac{b}{2m}t} e^{\pm i\omega_r t} \tag{3.35}$$

From Euler's identity the equation above can be written as

$$z = A_0 e^{-\frac{b}{2m}t}[\cos\, \omega_r t + i \sin\, \omega_r t] \tag{3.36}$$

The real part of the solution can be written in a general form as

$$z = A_0 e^{-\frac{b}{2m}t} \cos\, (\omega_r t + \varphi) \tag{3.37}$$

Importantly, equation (3.37) illustrates that the original amplitude A_0 decays with $-\frac{b}{2m}$. Therefore, this equation indicates that the energy of the oscillator is proportional to the square of the amplitude A_0 or simply proportional to A_0^2 and to the decaying expression $(e^{-\frac{b}{2m}t})^2$.

A typical definition follows from this derivation where τ is defined as

$$\frac{m}{b} = \tau \tag{3.38}$$

Since the units of m are kg and those of b are kg s^{-1}, τ is expressed in units of time. In particular, the definition of τ is the time required for the oscillator's energy to diminish by a factor of $\frac{1}{e}$. This can be shown as follows. From before (equation (3.8)), the total stored energy in an oscillator is

$$E = \frac{1}{2}kA^2 \tag{3.39}$$

Where, from equation (2.12), the following identity can be used to express k as

$$k = m\omega_0^2 \tag{3.40}$$

Combining equations (3.37)–(3.40)

$$E = \frac{1}{2}m\omega_0^2 A_0^2 e^{-\frac{t}{\tau}} \tag{3.41}$$

When $t = 0$,

$$E_0 = \frac{1}{2}m\omega_0^2 A_0^2 \tag{3.42}$$

When $t = \tau$,

$$E_\tau = \frac{1}{2}m\omega_0^2 e^{-1} \tag{3.43}$$

and this concludes the proof. The final expression for the energy stored in an oscillator can be expressed as

$$E = E_0 e^{-\frac{t}{\tau}} \tag{3.44}$$

The above result (equation (3.41)) assumes that there is dissipation and no drive (equation (2.5)) meaning that the amplitude must decay with time. The expression can also be found in other ways, for example, by means of the energy dissipation expressed as

$$P_{dis} = \frac{dE}{dt} = F_{dis}\dot{z} = -b\dot{z}\dot{z} = -b\dot{z}^2 \tag{3.45}$$

It is clear that P_{dis} is the instantaneous dissipation and for this reason it can be written in terms of the derivative of the total energy with time.

Using equation (3.10),

$$\langle KE \rangle \equiv \frac{1}{2}\left[\frac{1}{2}kA^2\right] = \frac{1}{2}E_T = \frac{1}{2}m\langle \dot{z} \rangle^2$$

$$E_T = m\langle \dot{z} \rangle^2 \tag{3.46}$$

where the term in brackets is the average velocity. From the above expression the average velocity can be expressed in terms of E, m (or k and ω_0). Combining equations (3.45) and (3.46) and assuming an average value for $\frac{dE}{dt}$(per cycle),

$$\frac{dE}{dt} = -b\langle \dot{z} \rangle^2 = -b\frac{E}{m} \tag{3.47}$$

From this expression the differential of E can be expressed as a function of time,

$$\frac{dE}{E} = \frac{-b}{m}dt \tag{3.48}$$

Integrating and using $\Delta t = (t - t_0)$

$$\int_{E_0}^{E} \frac{dE}{E} = \frac{-b}{m}\Delta t$$

$$\ln\left(\frac{E}{E_0}\right)\Bigg|_{E_0=E/t_0} = \frac{-b}{m}\Delta t\Bigg|_{t_0=0}$$

$$E = E_0 e^{-\frac{b}{m}\Delta t} \tag{3.49}$$

Finally, using the definition $\tau = \frac{m}{b}$ from equation (3.38) and the fact that $t_0 = 0$,

$$E = E_0 e^{-\frac{t}{\tau}} \tag{3.50}$$

This expression is equivalent to equation (3.41) and this completes the proof.

References

[1] Stroud K A and Booth D 2020 *Advanced Engineering Mathematics* (London: Red Globe Press)

[2] Alshehhi M, Alhassan S M and Chiesa M 2017 Dependence of surface aging on DNA topography investigated in attractive bimodal atomic force microscopy *Phys. Chem. Chem. Phys.* **19** 10231–6

[3] Santos S *et al* 2013 Stability, resolution, and ultra-low wear amplitude modulation atomic force microscopy of DNA: small amplitude small set-point imaging *Appl. Phys. Lett.* **103** 63702–5

[4] Amadei C A, Yang R, Chiesa M, Gleason K K and Santos S 2014 Revealing amphiphilic nanodomains of anti-biofouling polymer coatings *ACS Appl. Mater. Interfaces* **6** 4705–12

[5] Santos S, Olukan T A, Lai C-Y and Chiesa M 2021 *Hydration Dynamics and the Future of Small-Amplitude AFM Imaging in Air. Molecules.* (Switzerland: Basel) p 26

[6] Lai C-Y, Santos S and Chiesa M 2016 Systematic multidimensional quantification of nanoscale systems from bimodal atomic force microscopy data *ACS Nano* **10** 6265–72

[7] Lai C-Y, Olukan T, Santos S, Al Ghaferi A and Chiesa M 2015 The power laws of nanoscale forces under ambient conditions *Chem. Commun.* **51** 17619–22

[8] Chiesa M and Lai C-Y 2018 Surface aging investigation by means of an AFM-based methodology and the evolution of conservative nanoscale interactions *Phys. Chem. Chem. Phys.* **20** 19664–71

[9] Chiou Y-C *et al* 2018 Direct measurement of the magnitude of the van der Waals interaction of single and multilayer graphene *Langmuir* **34** 12335–43

[10] Amadei C A, Lai C-Y, Heskes D and Chiesa M 2014 Time dependent wettability of graphite upon ambient exposure: the role of water adsorption *J. Chem. Phys.* **141** 084709

[11] Lai C-Y *et al* 2014 A nanoscopic approach to studying evolution in graphene wettability *Carbon* **80** 784–92

[12] Amadei C A, Tang T C, Chiesa M and Santos S 2013 The aging of a surface and the evolution of conservative and dissipative nanoscale interactions *J. Chem. Phys.* **139** 084708

[13] Lai C-Y, Perri S, Santos S, Garcia R and Chiesa M 2016 Rapid quantitative chemical mapping of surfaces with sub-2 nm resolution *Nanoscale* **8** 9688–94

[14] Eichhorn A L and Dietz C 2021 Simultaneous deconvolution of in-plane and out-of-plane forces of hopg at the atomic scale under ambient conditions by multifrequency atomic force microscopy *Adv. Mater. Interfaces* **8** 2101288
[15] Santos S *et al* 2021 Investigating the ubiquitous presence of nanometric water films on surfaces *J. Phys. Chem.* C **125** 15759–72
[16] Wastl D S, Weymouth A J and Giessibl F J 2013 Optimizing atomic resolution of force microscopy in ambient conditions *Phys. Rev.* B **87** 245415
[17] Weymouth A J, Wastl D and Giessibl F J 2018 Advances in AFM: seeing atoms in ambient conditions *e-J. Surf. Sci. Nanotechnol.* **16** 351–5
[18] Giessibl F J 2003 Advances in atomic force microscopy *Rev. Mod. Phys.* **75** 949

Chapter 4

The driven oscillator

4.1 The linear form of the driven oscillator

The topic of driven oscillations is closely related to resonance and all the concepts discussed so far are exploited in its analysis. The general equation is equation (2.5), rewritten here for completeness as equation (4.1):

$$m\ddot{z} + \frac{m\omega_0}{Q}\dot{z} + kz = F_0 \cos \omega t + F_{ts} \tag{4.1}$$

The discussion in this chapter considers all the linear terms in the above expression, that is

$$m\ddot{z} + \frac{m\omega_0}{Q}\dot{z} + kz = F_0 \cos \omega t \tag{4.2}$$

The only term left out is F_{ts}, i.e., the nonlinear term. The second section of the book is dedicated to the analysis of equation (4.1), including F_{ts}. Since much of the nonlinear analysis is based on the results of the linear analysis the results of this section are very important. The rheological model for equation (4.2) is shown in figure 4.1.

All the terms on the left hand side of equation (4.2) have been discussed already. We recall that the addition of dissipation, the second term in equation (4.2), shifts the resonant frequency from ω_0 to ω_r where $\omega_r < \omega_0$. This however does not imply that ω_0 is redefined as ω_r. Rather, we still have the phenomena occurring at $\omega = \omega_0$ and the phenomena occurring at $\omega = \omega_r$. In particular, in the presence of dissipation the maximum amplitude A occurs at $\omega = \omega_r$ r. However, we still have that $\phi = 90$ when $\omega = \omega_0$.

It is necessary to understand that equation (4.2) produces two solutions, a steady-state solution, and a transient solution. The transient solution is the same as that already analysed via the concepts in section 2.3 and 3.3 resulting in equation (3.37)

$$z = A_0 e^{-\frac{t}{2\tau}} \cos\left(\omega_r' t + \varphi\right) \tag{4.3}$$

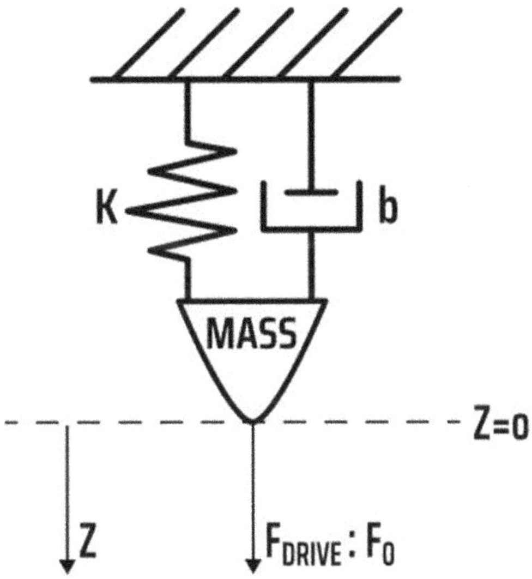

Figure 4.1. Rheological model corresponding to equation (4.2).

The transient solution is characterized by a resonant frequency that we term ω_r' in order to differentiate it from the resonant frequency that we will find in the presence of the driving force. Equation (4.3) is also characterized by a phase shift φ that indicates that for an arbitrary time t, the phase is φ.

Nevertheless, the transient and the steady-state solution 'coexist' until the amplitude in (4.3) vanishes compared to the amplitude given by the steady-state solution. Since the amplitude in is parametrized as a function of τ, the relevance of the transient solution, i.e., uncontrolled oscillation, depends on

$$z = A(t, \tau)\cos(\omega_r' t + \varphi) \tag{4.4}$$

From equation (3.38)

$$\tau = \frac{m}{b} = \frac{Q}{\omega_0} \tag{4.5}$$

The implication is that for a given ω_0, the larger the Q the longer the transient solution persists. In amplitude modulation (AM) AFM, the amplitude must be constant. The implication is that the minimum time for scanning controllably, or the maximum scan rate, is limited by Q. As stated, the less dissipation the larger the Q. Thus, while working in vacuum environments and low temperatures might be useful to minimize thermal noise, the value of Q is very high under such conditions. The implication is that, depending on the feedback, the scanning rate is further limited under vacuum. One must wait until the amplitude becomes approximately constant in order to operate the instrument in the amplitude modulation (AM) mode where the amplitude must remain constant. Otherwise noise will control the system and the image will be blurred. The opposite is true for liquid environments where Q is very low. Similar considerations can be drawn for frequency modulation (FM) where the frequency shift is to remain constant [1, 2]. These concepts will be discussed in the second section of the book. Thus, the time required for these transitory solutions (equation (4.4)) to vanish is crucial. Practically, this means that if the cantilever was driven really quickly, the oscillation amplitude would be dominated by the transient solutions, i.e., the dynamics of the cantilever would be controlled by the transient term $A_0 e^{-\frac{t}{2\tau}}$, making it impossible or difficult to maintain a constant amplitude. The purpose of the drive is to control and set a given amplitude as selected by the user. The solution controlled by the drive is that produced by the steady state. The object of this chapter is the steady-state solution.

In the steady state, the oscillation amplitude should be kept constant. This is particularly true in AM AFM where A is used as feedback. The physical implication is that, in the steady state, all of the energy that the drive delivers to the system is dissipated by the viscous term at a constant rate. First the drive builds up the amplitude and later such equilibrium is reached. In the presence of transients, the system will continue to dissipate energy and decrease A until the amplitude is constant. When the cantilever is driven far from the surface this amplitude is typically termed A_0.

In AM AFM the phase shift ϕ will produce contrast while the amplitude should be (ideally) flat, i.e., there should be no error in A as the amplitude should remain constant. The contrast channel A is thus typically termed the error channel in AM AFM. However, in frequency modulation (FM), amplitude contrast is possible (see section 4.2) since what remains constant is the frequency shift and A can freely vary.

Since equation (4.2) consists of linear terms only, the method employed in part 2 (equation (6.4)) to solve the equation is still valid. From there

$$z = Be^{i\omega t} \equiv x + iy \tag{4.6}$$

where if $z = x + iy$ is a solution of equation (2.17) so are x and y. Time derivatives produce the following result

$$z = Be^{i\omega t} \tag{4.7}$$

$$\dot{z} = i\omega Be^{i\omega t} \tag{4.8}$$

$$\ddot{z} = -\omega^2 Be^{i\omega t} \tag{4.9}$$

Substituting equations (4.7)–(4.9) into equation (4.2) gives

$$-m\omega^2 z + ib\omega z + m\omega_0^2 z = F_0 e^{i\omega t} \tag{4.10}$$

From equation (4.7)

$$e^{i\omega t} = \frac{z}{B} \tag{4.11}$$

Then, the drive can be written in its complex form as

$$e^{i\omega t} = \frac{z}{B} \tag{4.12}$$

Combining equations (4.11) and (4.12)

$$B = \frac{F_0}{m(\omega_0^2 - \omega^2) + ib\omega} \tag{4.13}$$

Using the basics of complex numbers and figure 4.2, the complex number z is

$$z = x + iy \tag{4.14}$$

can also be expressed in polar form

$$z = re^{i\phi} = \sqrt{x^2 + y^2}\, e^{i\phi}$$

$$\tan\phi = \frac{y}{x} \tag{4.15}$$

The form in equation (4.15), and particularly the equation of the tangent of the phase ϕ, are very important in oscillation theory. Then, equation (4.13) can be written in polar form as

$$B = \frac{F_0}{z} = \frac{F_0}{x + iy} \tag{4.16}$$

where

$$z = m(\omega_0^2 - \omega^2) + ib\omega = \sqrt{m^2(\omega_0^2 - \omega^2) + b^2\omega^2}\, e^{i\phi} \tag{4.17}$$

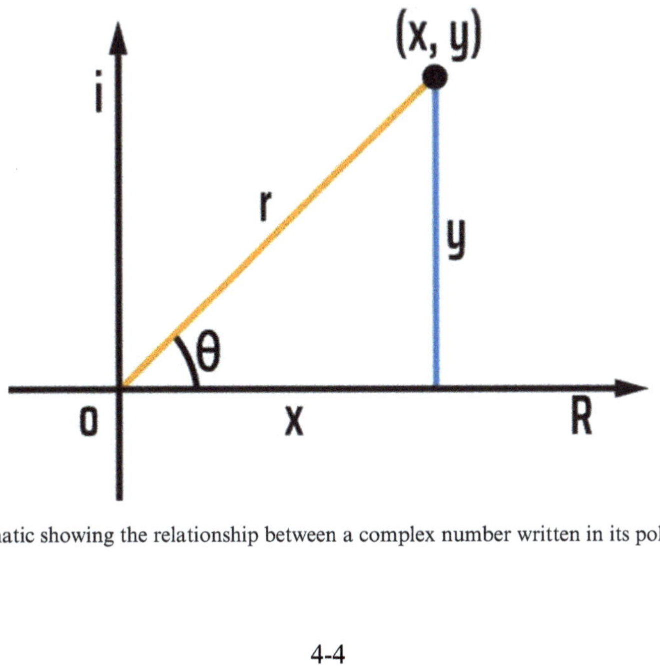

Figure 4.2. Schematic showing the relationship between a complex number written in its polar form and in its cartesian form.

and

$$\tan\phi = \frac{b\omega}{m(\omega_0^2 - \omega^2)} \tag{4.18}$$

Finally,

$$B = F_0 \frac{1}{\sqrt{m^2(\omega_0^2 - \omega^2) + b^2\omega^2}\, e^{i\phi}}$$

$$B = \frac{F_0 e^{-i\phi}}{\sqrt{m^2(\omega_0^2 - \omega^2) + b^2\omega^2}} \tag{4.19}$$

The above result implies that the complex solution of (4.2) is,

$$z = Be^{i\omega t} = Ae^{i(\omega t - \phi)} \tag{4.20}$$

where both amplitude A and phase ϕ are determined from known parameters as follows

$$A = \frac{F_0}{\sqrt{m^2(\omega_0^2 - \omega^2)^2 + b^2\omega^2}} \tag{4.21}$$

and

$$\tan\phi = \frac{b\omega}{m(\omega_0^2 - \omega^2)} \tag{4.22}$$

It is important to notice the negative sign in equation (4.20). This negative sign can be written there or directly in equation (4.22). Feynman emphasizes that the negative sign should be in equation (4.22) since this directly indicates that z lags the drive. The full complex solution is

$$z = Be^{i\omega t} = Ae^{i(\omega t - \phi)}$$

$$z = A\cos(\omega t - \phi) + iA\sin(\omega t - \phi) \tag{4.23}$$

It follows from linearity that both parts are solutions, meaning that a solution to the driven oscillator is

$$z = A\cos(\omega t - \phi) \tag{4.24}$$

where the phase shift ϕ, i.e., the lag between the response and the driving force, is given by equation (4.22). In particular, the response always lags the force as shown in figure 4.3. In the figure the drive force has been written in terms of the drive amplitude A_D which is not equivalent to the drive force F_0. In terms of A_D the drive force is

$$F_D = kA_D\cos(\omega t) \tag{4.25}$$

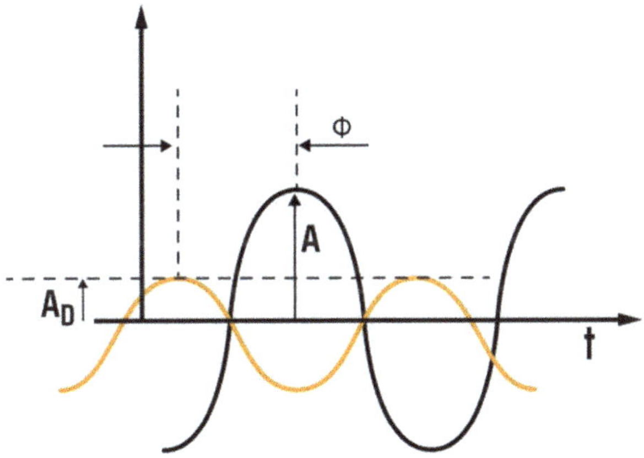

Figure 4.3. Illustration of the lagging of $z = A\cos(\omega t - \phi)$ with respect to $F_D = kA_D\cos(\omega t)$.

where k is the cantilever's spring constant, and the position of the drive is identified with $z_D = A_D\cos(\omega t)$. The rheological model discussing the drive force in terms of equation (4.25) is discussed in section 4.4. In particular equation (4.25) equivalent to

$$F_D = F_0\cos(\omega t) \tag{4.26}$$

The implication is that $F_0 = kA_D$. Anczykowski *et al* [3] elaborated a theory, i.e., rheological model and expressions (see section 4.4), where the interested reader can learn more about this approach. By comparing equations (4.25) or (4.26) to equation (4.24), the lagging of z with respect to the drive force can be plotted as a graph (figure 4.3).

4.2 Summary

As a main result, it is worth noting that all parameters in A (4.21) and ϕ (4.22) can be experimentally found when calibrating the cantilever in AFM. For example, figure 4.4 shows a standard frequency sweep acquired with a cypher AFM (Oxford Instruments) at a large enough distance from the surface so that $F_{ts} \to 0$. Under such conditions the linear equation (4.2) applies. It can be shown that [4] Q, ω_0, ω_r and k can be derived from such sweeps. This means that expressions can be found to relate frequency sweeps, or other experimental data or identities, to these parameters. Such expressions are exploited in AFM software. Some of the relationships have been developed so far in the chapters above.

For example, we have already shown how ω_r is obtained when there is dissipation, i.e., ω_r is that frequency which gives a maximum A in a frequency sweeps such as that in figure 4.4. In the next chapter we show how to relate ω_r to ω_o, so ω_0 can be indirectly computed from ω_r and Q (experimentally finding ω_r is easy since it is the ω of maximum amplitude in a sweep like that in figure 4.4). Note that another expression for ω_r was already found in section 2.3 when dissipation was allowed. The presence of a driving force however leads to another expression for ω_r. In the next chapter we also show how to compute Q directly from a frequency sweep such as

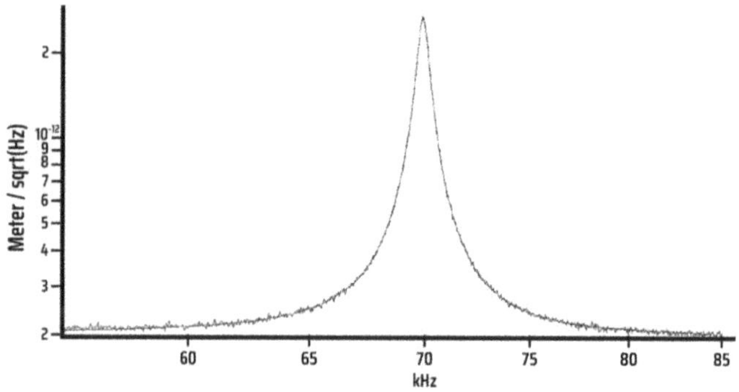

Figure 4.4. Standard frequency sweep in AFM where the resonance curve is seen to peak at $fr \approx 70$ kHz. The finite width of the curve indicates that there is dissipation. Thus Q takes on a finite value. Here the value of Q was approximately 100. The implication is that $\omega_r < \omega_0$. The phase shift is 90 at ω_0 while the amplitude A is maximum at ω_r.

that in figure 4.4. The spring constant k can be measured in many ways in AFM [4]. For example, looking at a force versus distance curve. Otherwise the cantilever manufacturer typically provides data on k.

4.3 The driven oscillator and resonance with a driving force

This section elaborates on the relationships between the Q factor, dissipation, ω_0, ω_r and driven response.

First, equation (4.21) provides the means to find the maximum value of A as a function of drive frequency ω. In AFM the drive frequency can be arbitrarily selected by the user. In particular, in AFM both F_0 and ω can be arbitrarily selected. This determines the drive force as observed from equation (4.26), i.e., $F_D = F_0 \cos(\omega t)$.

The amplitude is (equation (4.21)),

$$A = \frac{F_0}{\sqrt{m^2(\omega_0^2 - \omega^2)^2 + b^2\omega^2}}$$

Equating the derivative of the amplitude with respect to ω to zero gives the maximum amplitude $A_{\max}$

$$\frac{dA}{d\omega} = 0 \Rightarrow A_{\max}$$

$$\omega_r \mid_{A=A_{\max}} = \omega_0 \left[1 - \frac{1}{2Q^2}\right]^{1/2} \tag{4.27}$$

The value $A_{\max}$ determines ω_r. Again $\omega_r < \omega_0$. This resonance is different from that found in the presence of dissipation and the absence of drive (compare with equation (6.17)). This result can be obtained algebraically, but it is easier to confirm it by exploiting computing software such as Matlab. The script below provides the result in equation (4.27). In the script w_ratio=ω/ω₀ has been introduced to simplify equation (4.21).

```
Matlab script to solve equation (4.21) and produce (4.27).
clear all
clc
syms Q w_ratio
y1= ((1-(w_ratio)^2)^2+(w_ratio/Q)^2)^0.5;
diff_w = diff(y1, w_ratio)
solve_w = solve(diff_w ==0, w_ratio)
```

Figure 4.5 shows a relevant part of a frequency sweep obtained by plotting equation (4.21). In the vertical axis, A is plotted, and in the horizontal axis ω is plotted normalized in terms of ω_0, i.e., $\frac{\omega}{\omega_0}$. Only the region for $0.9 < \frac{\omega}{\omega_0} < 1.1$ is shown in order to showcase that the maximum amplitude A does not result in $\frac{\omega}{\omega_0} = 1$. This is in agreement with equation (4.27). The phase shift ϕ however is still 90° where $\omega = \omega_0$. This can be confirmed from equation (4.22). The curve is that of the linear response from equation (4.2).

There is a way to derive an expression for the Q factor in terms of a frequency sweep as shown in figure 4.5. A schematic showing relevant parameters for such derivation is shown in figure 4.6. The first thing to note is that for the derivation to hold $\omega_r \approx \omega_0$. This happens when Q is large enough. The larger Q is, the better the following derivation holds. The amplitude A in figure 4.6 is parametrized by ω, i.e.,

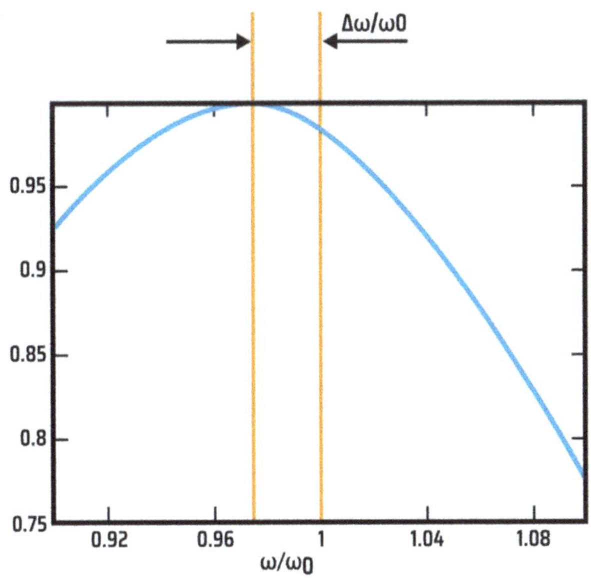

Figure 4.5. Part of a frequency sweep obtained by plotting equation (4.21) where A is plotted against the normalized frequency $\frac{\omega}{\omega_0}$. The maximum value of A, i.e., A_{max}, does not coincide with ω_0. Rather A_{max} occurs when $\frac{\omega}{\omega_0} < 0$. The actual value of ω_r is found from equation (4.27).

the drive frequency. Then, from equation (4.21) and since the power of the system is proportional to the square of the amplitude (equation (3.8))

$$A^2 = \frac{F_0^2}{m^2(\omega_0^2 - \omega^2)^2 + b^2\omega^2} \tag{4.28}$$

In his lectures on resonance, Feynman works also with the square of the amplitude since the derivation of Q follows from the A^2 versus ω curve. Feynman further proposed to the readers to find the relationship between $\Delta\omega$ as shown in figure 4.6. This solution must be found in terms of the square of the amplitude, i.e., the term proportional to power or stored energy.

Here, such a relationship is derived as follows. If we term the maximum amplitude squared A_0^2, the objective is to find $A^2 = \frac{1}{2}A_0^2$ analytically. This will relate $\Delta\omega$ to ω_0 and Q. It follows that some assumptions make this relatively easy. First, following the advice of Feynman in his lectures, we assume that $\omega_r \cong \omega_0$ (this assumption is key since the relationship to be found will hold only for large values of Q). It follows that

$$\omega_0^2 - \omega^2 \approx (\omega_0 - \omega)2\omega_0 \tag{4.29}$$

Then,

$$A^2 = \frac{F_0^2}{m^2(\omega_0^2 - \omega^2)^2 + b^2\omega^2} \approx \frac{F_0^2}{4\omega_0^2 m^2(\omega_0 - \omega)^2 + b^2\omega^2} \tag{4.30}$$

The following identity can also be employed to simplify the equation and write equation (4.30) in term of Q

$$b^2\omega^2 = \frac{m^2\omega_0^2\omega^2}{Q^2} \tag{4.31}$$

We also recall that, by definition,

$$\gamma = \frac{\omega_0}{Q} \tag{4.32}$$

Combining equations (4.30) and (4.31) the general expression for the amplitude as a function of ω is

$$A^2 \approx \frac{F_0^2}{4\omega_0^2 m^2\left[(\omega_0 - \omega)^2 + \left[\frac{\omega}{2Q}\right]^2\right]}$$

$$A^2 \approx \frac{F_0^2}{4\omega_0^2 m^2\left[(\omega_0 - \omega)^2 + \left[\frac{\gamma}{2}\right]^2\right]} \tag{4.33}$$

In equation (4.33) it has been assumed that when $\omega \approx \omega_0$, the width of the curve in figure 4.6 is $\omega = \omega_0 + \frac{\Delta\omega}{2}$ where $\gamma = \frac{\omega_0}{Q} \approx \frac{\omega}{Q}$. This last assumption is key and holds only for large Q.

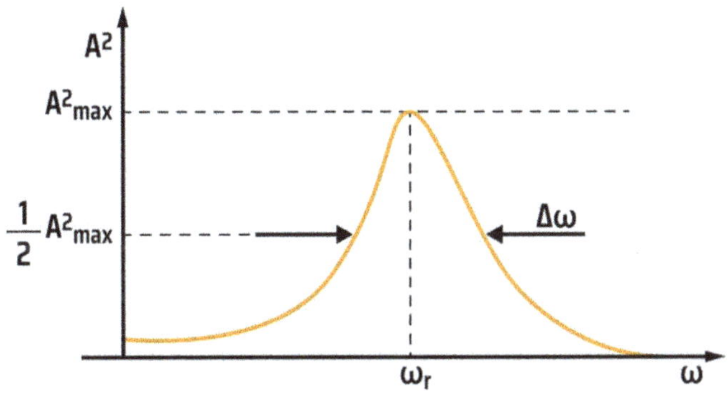

Figure 4.6. Amplitude squared A^2 versus frequency ω where the width $\Delta\omega$ at half power is shown.

With the above assumption equation (4.33) can be further simplified

$$A^2 \approx \frac{F_0^2}{4k^2\left[\left(1 - \frac{\omega}{\omega_0}\right)^2 + \left[\frac{1}{2Q}\right]^2\right]} \tag{4.34}$$

The maximum stored energy occurs at maximum amplitude A_0, or stored energy, which is proportional to the square of the amplitude. This has been shown to result (equation (4.27)) when $\omega = \omega_r$ where $\omega_r \approx \omega_0$ if Q is large enough. Then

$$A_0^2 \approx \frac{F_0^2}{4k^2\left[\frac{1}{2Q}\right]^2}$$

$$A_0^2 \approx \left[\frac{F_0 Q}{k}\right]^2 \tag{4.35}$$

The above (equation (4.35)) is a typical result employed in AFM to compute the drive force F_0 in Newtons in terms of the observables A_0, k, and Q when working at or near resonance. The square of the amplitude at half maximum occurs when $\Delta\omega = 2\,|(\omega_0 - \omega)|$ as illustrated in figure 4.6. The square of the amplitude at $\left[\frac{1}{2\Delta\omega A_0}\right]^2$ can be found from equation (4.34)

$$A^2(\omega = \omega_0 \pm \Delta\omega) \approx \frac{F_0^2}{4k^2\left[\left(1 - \frac{\omega}{\omega_0}\right)^2 + \left[\frac{1}{2Q}\right]^2\right]}$$

$$A^2(\omega = \omega_0 \pm \Delta\omega) \approx \frac{F_0^2}{4k^2\left[\left(1 - \frac{\omega_0 \pm \Delta\omega/2}{\omega_0}\right)^2 + \left[\frac{1}{2Q}\right]^2\right]}$$

$$A^2(\omega = \omega_0 \pm \Delta\omega) \approx \frac{F_0^2}{4k^2\left[\left(\frac{\pm\Delta\omega}{2\omega_0}\right)^2 + \left[\frac{1}{2Q}\right]^2\right]} \tag{4.36}$$

Since the drive frequency at maximum amplitude can be written as $\omega = \omega_0 \pm \frac{\Delta\omega}{2}$. Combining the constraint $A^2 = \frac{1}{2}A_0^2$ and equations (4.35) and (4.36)

$$\frac{1}{2}\left[\frac{F_0 Q}{k}\right]^2 = \frac{F_0^2}{4k^2\left[\left(\frac{\pm\Delta\omega}{2\omega_0}\right)^2 + \left[\frac{1}{2Q}\right]^2\right]}$$

$$2Q^2 = \frac{1}{\left[\left(\frac{\pm\Delta\omega}{2\omega_0}\right)^2 + \left[\frac{1}{2Q}\right]^2\right]}$$

$$\left(\frac{\pm\Delta\omega}{2\omega_0}\right)^2 2Q^2 + \frac{2Q^2}{[2Q]^2} = 1$$

$$\left(\frac{\pm\Delta\omega}{2\omega_0}\right)^2 2Q^2 + \frac{1}{2} = 1$$

$$\left(\frac{\pm\Delta\omega}{2\omega_0}\right)^2 = \frac{1}{[2Q]^2} \tag{4.37}$$

Equation (4.37) shows that the Q factor is related to the width of the resonance curve $\Delta\omega$ (figure 4.6) as follows

$$\frac{\Delta\omega}{\omega_0} = \frac{1}{Q} \tag{4.38}$$

It is interesting to see how other interesting relationships follow from equation (4.38)

$$\frac{\Delta\omega}{\omega_0} \approx \frac{1}{Q} \approx \frac{\Delta f}{f_0}$$

$$\Delta\omega = \frac{\omega_0}{Q} \equiv \gamma$$

$$\Delta\omega Q = \omega_0 \tag{4.39}$$

4.4 Other derivations of Q and energy considerations in the linear model

The work done by the drive in the steady state must be equal to the energy dissipated via the viscous term contained in the equation describing the driven oscillator (equation (4.2))

$$F_D = F_0 \cos \omega t$$

$$m\ddot{z} + \frac{m\omega_0}{Q}\dot{z} + kz = F_0 \cos \omega t \tag{4.40}$$

The work done by the drive is expressed in its differential form as

$$dz = \dot{z}dt$$

$$dW = F_D dz \tag{4.41}$$

Where, from equation (4.24)

$$\dot{z} = -\omega A \sin\,(\omega t - \phi) \tag{4.42}$$

Finally, the work done per cycle results from combining equations (4.40)–(4.42)

$$W_D = \int_0^T F_D \cdot \dot{z}dt$$

$$W_D = -F_0 A\omega \int_0^T \sin\,(\omega t - \phi)\cos \omega t dt$$

$$W_D = -F_0 A\omega \int_0^T (\sin\,(\omega t) \cos \phi - \cos\,(\omega t) \sin \phi)\cos \omega t dt$$

$$W_D = F_0 A\omega \left[\int_0^T \cos^2\,(\omega t)dt \right] \cdot \sin \phi \tag{4.43}$$

The above results from the orthogonality between the cosine and the sine functions. Solving equation (4.43), the work done per cycle by the drive is

$$W_D = F_0 A\omega \left[\frac{T}{2} \right] \sin \phi$$

$$T = \frac{2\pi}{\omega}$$

$$W_D = \pi F_0 A \sin \phi \tag{4.44}$$

The physical interpretation of equation (4.44) is that the maximum work done per cycle by the drive occurs approximately when $\sin \phi = 1$, namely, when, from equation (4.22), $\omega = \omega_0$. This definition is approximate only because W_D multiplies A and $\sin \phi$ and A is not maximum when $\sin \phi = 1$. The implication is that $\sin \phi = 1$ does not coincide with the value $\omega = \omega_r$ that produces the maximum amplitude either. The implication is that the phase shift is approximately 90° for the maximum work done by the drive. Importantly, it will be shown in section 4.2 that the same expression is found when computing the energy dissipated by the nonlinear term F_{ts}.

The differential of the energy dissipated dE by the viscous term in equation (4.40) is given by the differential equation

$$dz = \dot{z}dt$$

$$dE = F_b \cdot dz \tag{4.45}$$

Then, the energy dissipated per cycle by the viscous force follows by solving the following integral

$$E = \int_0^T F_b \cdot \dot{z}dt$$

$$E = \int_0^T \frac{m\omega_0}{Q}\dot{z} \cdot \dot{z}dt$$

$$E = \frac{m\omega_0}{Q}\int_0^T [\omega A \sin(\omega t - \phi)]^2 \, dt$$

$$E = \frac{m\omega_0}{Q}\omega^2 A^2 \int_0^T [\sin(\omega t)\cos\phi - \cos(\omega t)\sin\phi]^2 \, dt$$

$$E = \frac{m\omega_0}{Q}\omega^2 A^2 \int_0^T \{[\sin(\omega t)\cos\phi]^2 + [\cos(\omega t)\sin\phi]^2 - 2\sin(\omega t)\cos(\omega t)\sin\phi\cos\phi\}dt$$

$$E = -\frac{m\omega_0}{Q}\omega^2 A^2 \int_0^T \{[\sin(\omega t)\cos\phi]^2 + [\cos(\omega t)\sin\phi]^2\}dt$$

$$E = -\frac{m\omega_0}{Q}\omega^2 A^2 \int_0^T \{[\sin(\omega t)]^2 \cos^2\phi + [\cos(\omega t)]^2 \sin^2\phi\}dt$$

$$E = -\frac{m\omega_0}{Q}\omega^2 A^2 \left\{\frac{T}{2}\cos^2\phi + \frac{T}{2}\sin^2\phi\right\}$$

$$E = -\frac{m\omega_0}{Q}\omega^2 A^2 \frac{T}{2}[\cos^2\phi + \sin^2\phi]$$

$$E = -\frac{m\omega_0}{Q}\omega A^2 \pi \tag{4.46}$$

The Q factor can now be expressed as a function of energy dissipation per cycle normalized by the total energy. Equation (4.46) is also equivalent to that found when developing the nonlinear theory equation (6.24). The total stored energy is E_T (equation (3.8))

$$E_T = \frac{1}{2}kA^2$$

The ratio between equation (4.46) and equation (3.8) is, by assuming $\omega \approx \omega_0$,

$$\frac{|[E]_{\text{cycle}}|}{E_T} = \frac{\frac{k}{Q}A^2\pi}{\frac{1}{2}kA^2}$$

$$\frac{|[E]_{\text{cycle}}|}{E_T} = \frac{2\pi}{Q} \tag{4.47}$$

It follows that the energy dissipated per cycle by the viscous term, or equivalently (see equation (4.49) and discussion), the work done by the drive per cycle, is 2π divided by Q. As expected, the larger the Q the smaller this ratio. The implication is that the larger the Q factor the longer it takes for the system to lose energy via the viscous term.

Finally, the average power dissipated per cycle is simply equation (4.46) divided by T

$$\langle P \rangle = \frac{E}{T} = \frac{1}{2}\frac{m\omega_0}{Q}\omega^2 A^2 \tag{4.48}$$

If, as in the case of the driven oscillator in the steady state, all the power of the drive is dissipated by the dissipative term, equation (4.48) should be the same as equation (4.40). This can be shown as follows:

$$\langle P \rangle = \frac{W_D}{T} = \frac{\pi F_0 A \sin\phi}{T}$$

$$\langle P \rangle = \frac{1}{2} F_0 \omega A \sin\phi \tag{4.49}$$

Looking at the energy delivered by the drive instead (equation (4.44)), and if in the steady state all the energy dissipated per cycle must be delivered by the drive, it follows that

$$W_D = \pi F_0 A \sin\phi$$

$$\frac{|\,[W_D]_{\text{cycle}}\,|}{E_T} = \frac{\pi F_0 A \sin\phi}{\frac{1}{2}kA^2} \tag{4.50}$$

If $\omega \approx \omega_0$, it follows from the general expression for A as a function of ω for the driven oscillator (equation (4.21)) that

$$F_0 = \frac{kA}{Q} \tag{4.51}$$

Combining equations (4.50) and (4.51)

$$\frac{[W_D]_{\text{cycle}}}{E_T} = \frac{\pi \frac{kA}{Q} A \sin\phi}{\frac{1}{2}kA^2}$$

$$\frac{[W_D]_{\text{cycle}}}{E_T} = \frac{2\pi}{Q}\sin\phi \tag{4.52}$$

Since $\omega \approx \omega_0$, $\sin\phi = 1$. This follows from equation (4.22). Then

$$\frac{[W_D]_{\text{cycle}}}{E_T} = \frac{2\pi}{Q} \tag{4.53}$$

As expected, equation (4.53) is equivalent to equation (4.47) when $\sin\phi = 1 (\omega \approx \omega_0)$

From equation (4.37) a relationship with $\Delta\omega$, or the width of the curve (figure 4.6), can be found

$$\frac{[W_D]_{\text{cycle}}}{E_T} = 2\pi\frac{\Delta\omega}{\omega_0} \tag{4.54}$$

It follows that the energy dissipated per cycle can be easily found from the geometry of the frequency sweep (see figure 4.6). This result is very useful since it allows us to determine the Q factor, the power of the drive and the energy dissipated by the viscous term by inspection of a frequency sweep.

4.5 Another way of looking at the drive

As stated, some authors [3] express the drive as k_{z_D}; see equation (4.25). Then the position of the drive is

$$z_D = A_D \cos\,(\omega t) \tag{4.55}$$

The drive force is derived from the rheological model in figure 4.7

$$F_D = k[z(t) - z_D(t)] \tag{4.56}$$

where, as usual, the position of the point-mass is

$$z(t) = A\cos(\omega t - \phi) \tag{4.57}$$

The power delivered by the drive per cycle can now be written as follows

$$\langle P_D\rangle = \frac{1}{T}\int_0^T F_D\dot{z}_D dt \tag{4.58}$$

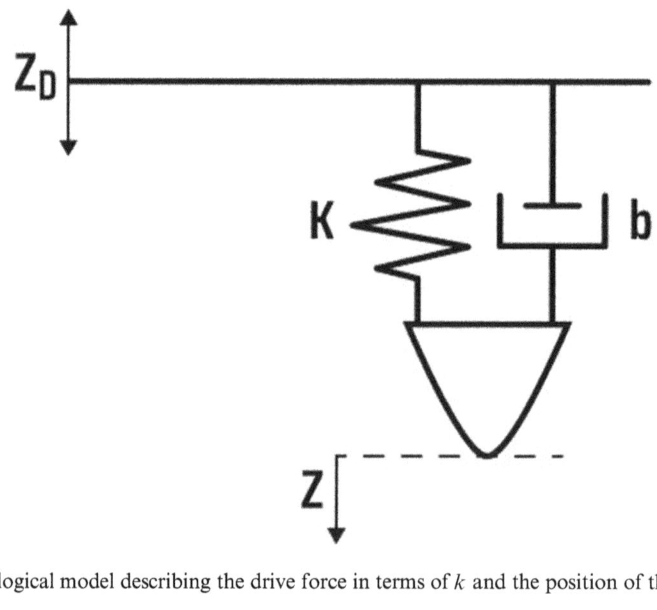

Figure 4.7. Rheological model describing the drive force in terms of k and the position of the drive z_D relative to the mass.

Combining equations (4.55)–(4.58)

$$\langle P_D \rangle = -\frac{1}{T}\int_0^T k\left[z(t) - z_D(t)\right]\dot{z}_D dt$$

$$\langle P_D \rangle = -\frac{1}{T}\int_0^T k\left[A\cos(\omega t - \phi) - A_D\cos(\omega t)\right] - A_D\omega \sin(\omega t)dt$$

$$\langle P_D \rangle = -\frac{k}{T}\int_0^T - AA_D\omega\cos(\omega t - \phi)\sin(\omega t) + \omega A_D^2\cos(\omega t)\sin(\omega t)dt$$

$$\langle P_D \rangle = \frac{k}{T}\int_0^T AA_D\omega\cos(\omega t - \phi)\sin(\omega t)dt \tag{4.59}$$

where, from orthogonality, we obtained the result in equation (4.59). Solving the integral

$$\langle P_D \rangle = \frac{AA_Dk\omega}{T}\sin(\phi)\int_0^T \sin^2(\omega t)dt$$

$$\langle P_D \rangle = \frac{AA_Dk\omega}{T}\frac{T}{2}\sin(\phi)$$

$$\langle P_D \rangle = \frac{1}{2}AA_Dk\omega\sin(\phi) \tag{4.60}$$

The power delivered (equation (4.46)) by the drive when modelling the drive as shown in equation (4.2) should be the same as that obtained in equation (4.60). In short

$$F_D = F_0\cos\omega t$$

$$\langle P_D \rangle = \frac{1}{2}F_0\omega A\sin\phi \tag{4.61}$$

Combining equations (4.60) and (4.61)

$$\frac{1}{2}AA_Dk\omega\sin(\phi) = \frac{1}{2}AF_0\omega\sin\phi$$

$$F_0 = kA_D \tag{4.62}$$

This concludes the proof. The above expression, equation (4.62), can be employed to relate the drive amplitude A_D to the drive force F_0. Combining equations (4.62) and (4.21)

$$A = \frac{F_0}{\sqrt{m^2(\omega_0^2 - \omega^2)^2 + b^2\omega^2}}$$

$$F_0 = kA_D$$

$$F_0 = A\sqrt{m^2(\omega_0^2 - \omega^2)^2 + b^2\omega^2}$$

$$\sqrt{m^2(\omega_0^2 - \omega^2)^2 + b^2\omega^2} = kA_D \tag{4.63}$$

The above expression can be simplified by using $k = m\omega_0^2$

$$kA_D = Am\sqrt{(\omega_0^2 - \omega^2)^2 + \frac{\omega_0^2\omega^2}{Q^2}}$$

$$kA_D = kA\sqrt{\left(1 - \frac{\omega^2}{\omega_0^2}\right)^2 + \frac{\omega^2}{\omega_0^2 Q^2}}$$

$$A_D = A\sqrt{\left(1 - \frac{\omega^2}{\omega_0^2}\right)^2 + \frac{\omega^2}{\omega_0^2 Q^2}} \tag{4.64}$$

The expression in equation (4.64) is important in that it relates A_D with A in terms that can be found by inspecting a frequency sweep. In short, since Q, A and ω_0 can be experimentally, and easily, found, A_D can be easily determined indirectly. The main body of this book has been, so far, to determine these parameters, i.e., Q, ω_r, etc, from experimental observables. The drive frequency ω is easily set and known in AFM.

$$\omega = \omega_0$$

Finally, at

$$A_D = A\sqrt{\frac{1}{Q^2}}$$

$$A_D = \frac{A}{Q} \tag{4.65}$$

As a final note, in dynamic AFM, where the cantilever is driven analogously to the driven oscillator discussed here, and when the cantilever is driven far from the surface, the linear equation (equation (4.2)) approximately applies. Then if we term the amplitude far from the surface the free amplitude A_0, i.e., the amplitude of oscillation in the absence of tip-sample force, the expression for A_D directly follows from observables

$$A_D = \frac{A_0}{Q} \tag{4.66}$$

The above expression is typically used in dynamic AFM, and, as noted, this expression is derived from the linear theory analysis in section 1 (chapters 1–4) of this book.

4.6 Summary of the chapter

The linear theory discussed in this section is very important both to the AFM field and to many other much broader fields. For this reason, whatever can be learnt from the liner theory can be easily applied across fields. Furthermore, there is always something new to learn about oscillation theory. As an example, the reader can check a 2022 paper where oscillators are employed to model complex phenomena such as brain activity in the field of functional Magnetic Resonance Imaging (fMRI). In short, the concepts employed in fMRI are the same as those discussed through chapters 1–4 here. For example, Bauer *et al* claim, of their fMRI model, that 'This is a system of ordinary differential equations (ODEs) describing the temporal change of the phases $\varphi 1$, … , φ_r of r oscillators, which are coupled by the sine of their phase differences' [5].

In summary, the reader will do well to go through the expressions discussed in this chapter with the aim to understand the physical interpretation, the limitations of the assumptions and possible applications in a variety of fields. The next section deals with the nonlinear formalism.

References

[1] García R and Perez R 2002 Dynamic atomic force microscopy methods *Surf. Sci. Rep.* **47** 197–301

[2] Giessibl F J 2003 Advances in atomic force microscopy *Rev. Mod. Phys.* **75** 949

[3] Anczykowski B, Gotsmann B, Fuchs H, Cleveland J P and Elings V B 1999 How to measure energy dissipation in dynamic mode atomic force microscopy *Appl. Surf. Sci.* **140** 376–82

[4] Sader J E *et al* 2012 Spring constant calibration of atomic force microscope cantilevers of arbitrary shape *Rev. Sci. Instrum.* **83** 103705–16

[5] Bauer L G *et al* 2022 Quantification of Kuramoto coupling between intrinsic brain networks applied to FMRI data in major depressive disorder *Front. Comput. Neurosci.* **16** 729556

Part II

Nonlinear theory and applications in AFM

IOP Publishing

Oscillations

Theory and applications in AFM

Tuza Adeyemi Olukan, Sergio Santos, Lamiaa Sami Elsherbiny and Matteo Chiesa

Chapter 5

Nonlinearities and the driven oscillator

5.1 Linear differential equations

This is the second part of the book on oscillations. The aim is to expand the linear theory of oscillations with a focus on the point-mass model. The point-mass model is an ideal system, but it has many applications in science since many systems can be expressed as a point mass attached to a spring. For example, Wilhelm H Westphal writes that '[A] point mass is an idealization of a real solid body. It possesses mass, but its dimensions are assumed to be so small that its location can be sufficiently accurately defined by the position of a point' [1].

For nonlinearity we understand the standard mathematical definition, namely, a system for which the output is not proportional to its input. The Wikipedia entry [2] on nonlinear systems opens with two interesting statements. First,

(1) 'In mathematics and science, a nonlinear system is a system in which the change of the output is not proportional to the change of the input.'

It is typical for people to think in linear terms, that is, proportionally. For example, if you work twice the hours it could be expected that you make twice the money. If something has twice the mass, it should weigh twice as much, etc. Many times however, whatever multiplies a variable does not do so proportionally to the variable.

The second statement in the Wikipedia entry reads,

(2) 'Nonlinear problems are of interest to engineers, biologists, physicists, mathematicians, and many other scientists because most systems are inherently nonlinear in nature.'

This second statement in the entry shows that, while linearity is relevant and important, the nonlinear analysis is many times more accurate in its representation of reality.

While the above definitions are useful, it is important to give a strict definition of nonlinearity since, here, the theme will be discussed both theoretically and

doi:10.1088/978-0-7503-5809-5ch5

practically. The theme of oscillations is further discussed in this text by working on the equations of motion. These are differential equations. In this respect, Feynman more precisely defines what a linear differential equation is, namely, '[A] linear differential equation is a differential equation consisting of a sum of several terms, each term being a derivative of the dependent variable with respect to the independent variable, and multiplied by some constant'. The above statement is translated into a linear differential equation with constant coefficients as follows:

$$c_k \frac{d^k x}{dt^k} + c_{k-1} \frac{d^{k-1} x}{dt^{k-1}} + \cdots + c_1 \frac{dx}{dt} + c_o x = f(t) \tag{5.1}$$

The dependent variable in equation (5.1) x and the independent variable is t. The coefficients c_k are constants. Physically the independent variable t might stand for time, but it does not have to. Furthermore, the function $f(t)$ might have any functional relationship with t with the only limitation being, for linearity to apply, that x is not involved. If x was involved in a way that the function $f(t)$ can be assimilated as one of the linear terms on the left of equation (5.1), the equation is still linear. The coefficients however might not be constant if they depend on t. If x appears in ways different to those of the linear terms on the left of equation (5.1), the equation is nonlinear. The tip-sample force in AFM is typically nonlinear implying that the term on the right of equation (5.1) is nonlinear. It is not coincidental that Richard Feynman starts his discussion on oscillations with a chapter on linear differential equations and the methods to solve these equations.

Everything said about the linear system in section 5.1 will be assumed here. The results from those chapters will be explored here for the analysis of the nonlinear system when possible. The linear system is expressed in equation (5.2) while the nonlinear system is expressed in equation (5.3). F_{ts} means tip-sample force and it is typical terminology in AFM, but the term stands for any nonlinear force in general. If the term represents other than force, the solution of the equations is equivalent, only the physical interpretation chances according to the phenomena represented.

$$m\frac{d^2 z}{dt^2} + \frac{m\omega_0}{Q}\frac{dz}{dt} + kz = F_0 \cos \omega t \tag{5.2}$$

$$m\frac{d^2 z}{dt^2} + \frac{m\omega_0}{Q}\frac{dz}{dt} + kz = F_{ts} + F_0 \cos \omega t \tag{5.3}$$

where all the terms in the equations are defined in the glossary and in the previous chapters. Equation (5.2) has been extensively explored in section 5.1. The objective of this section is the analysis of equation (5.3). The rheological models for both equations are shown in figure 2.1. As stated, the linear model can be directly applied to many systems of interest. For example, it turns out that the linear model for the oscillator, the phenomena related to 'an electron orbiting a massive, stationary nucleus'[1], can be modelled as a point-mass model on spring with a driven sinusoid.

[1] Wikipedia, Lorentz Oscillator Model. *https://en.wikipedia.org/wiki/Lorentz_oscillator_model* 2022.

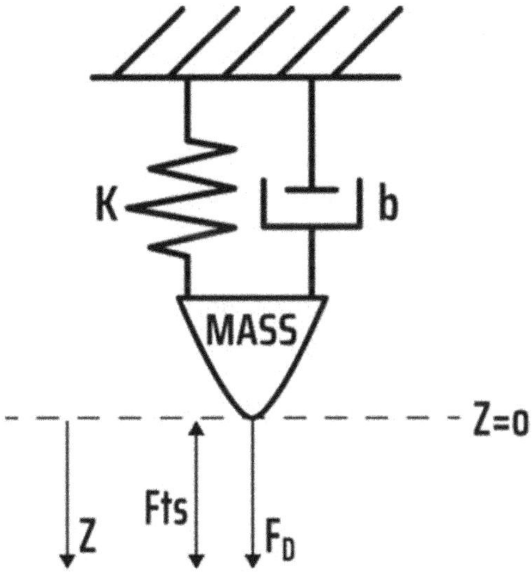

Figure 5.1. Rheological model representing the tip's motion and forces in the equation of motion in equation (5.3).

This model is termed 'the Lorentz oscillator model'. For example, Colton says that '[T]he Lorentz oscillator model, also known as the Drude–Lorentz oscillator model, involves modelling an electron as a driven damped harmonic oscillator' [3]. Levi says '[T]he Lorentz oscillator model applies the classical concepts of a driven damped mechanical oscillator to the problem of an electromagnetic field interacting with a dielectric material' [4]. Almog *et al* have a great document on the applications of this model [5]. Since the Lorentz oscillator model is based on phenomena that can be represented by a differential equation that is equivalent to the one discussed here (part I) for the AFM system, the resulting solutions and interpretations are also the same.

The nonlinear formalism results from the addition of the tip-sample force F_{ts} in equation (5.3) as shown in the rheological model in figure 5.1. F_0 is the magnitude of the driving force F_{DRIVE} or F_D.

5.2 Nonlinear differential equation of motion for the driven oscillator

It is important to recognize from the beginning that the nonlinear results can be derived from the linear ones by considering 'effective' parameters. This concept is not new or unique to the nonlinear analysis. For example, when the mass m is considered in the harmonic oscillator, the value of m is 'effective' in that it is the value m necessary for the model to adjust to, or better represent, the real system. The implication is that the actual mass m of the system does not need to coincide with the effective mass m. Many times, when speaking of effective values people use the prime notation, i.e., m', or simply write as a subscript eff for effective, i.e., m_{eff}.

The concept 'effective' will be exploited here to derive an 'effective' Q factor and an 'effective' resonance frequency ω_r when dealing with the nonlinear system. Effective will mean that were we use effective values for Q and ω_r, the interpretation of the nonlinear system is reduced to the interpretation and terminology used in the linear analysis. Of course, the interpretation and discussion must be considered with care. To provide such a discussion is also the focus of this book. In summary, most of the linear results can be extrapolated or transformed so they can be adapted to the analysis of the nonlinear system.

If F_{ts} is negligible compared to every other term in equation (5.3), F_{ts} has no significant effect in the response. In such case, the system can be analysed using the linear theory derived from the analysis of equation (5.2) only (section 5.1). However, as F_{ts} becomes comparable to the other terms, the nonlinear phenomena must be considered.

To begin, the comparison between the following equations can be considered,

$$m\frac{d^2z}{dt^2} + \frac{m\omega_0}{Q}\frac{dz}{dt} + kz = F_0 \cos \omega t \quad \text{linear} \tag{5.4}$$

$$m\frac{d^2z}{dt^2} + \frac{m\omega_0}{Q}\frac{dz}{dt} + kz = F_0 \cos \omega t - \frac{\alpha}{z^2} \quad z > B, \text{ nonlinear attractive} \tag{5.5}$$

$$m\frac{d^2z}{dt^2} + \frac{m\omega_0}{Q}\frac{dz}{dt} + kz = F_0 \cos \omega t + \beta z^{3/2} \quad z < C, \text{ nonlinear repulsive} \tag{5.6}$$

$$\begin{aligned} &m\frac{d^2z}{dt^2} + \frac{m\omega_0}{Q}\frac{dz}{dt} + kz = \\ &F_0 \cos \omega t - \frac{\alpha}{z^2} + \beta \mid z \mid^{3/2} \quad \text{nonlinear attractive–repulsive} \end{aligned} \tag{5.7}$$

Figures 5.2–5.5 show the difference between the linear and the nonlinear response in terms of the presence or absence of a nonlinear force. The response is typically represented by the amplitude and phase shift ϕ as a function of drive frequency in equations (5.4)–(5.7) as prescribed by $F_0 \cos \omega t$. Four cases will be discussed next from the linear response to cases where the nonlinear term is 'attractive', 'repulsive', and a combination of both. These cases are important in AFM because atomic forces behave in a similar way. Arvind Raman has a similar discussion in 'nanoHUB-U Fundamentals of AFM L2.3: Analytical Theory—Nonlinearity, Virial, and Dissipation' [6].

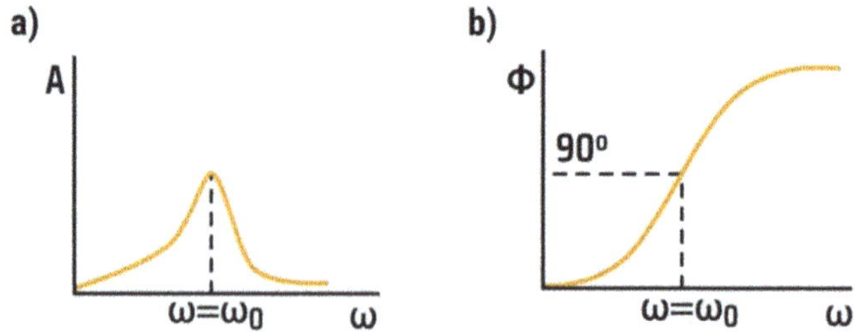

Figure 5.2. Linear case. Illustration of a standard frequency sweep showing the (a) amplitude A and (b) phase ϕ response where the resonance curve is seen to peak at $\omega_r \approx \omega_0$. The finite width of the curve indicates that there is dissipation meaning that Q takes on a finite value. The shapes in (a) and (b) are standard for a driven oscillator and are well represented by a linear model as that in equation (5.2). The equations for A and ϕ (equations (4.21) and (4.22)) are solved analytically from (5.2).

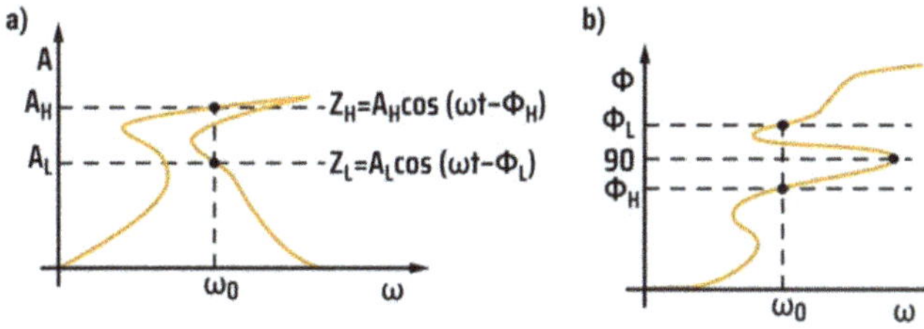

Figure 5.3. Nonlinear case. Illustration of a standard frequency sweep showing the (a) amplitude A and (b) phase ϕ response. The curves show distortions due to the presence of nonlinear attractive and repulsive forces. The distortions are large at and near ω_0 where two stables branches are available, i.e. H and L (high and low state). Far from ω_0 the oscillator behaves as in the linear case.

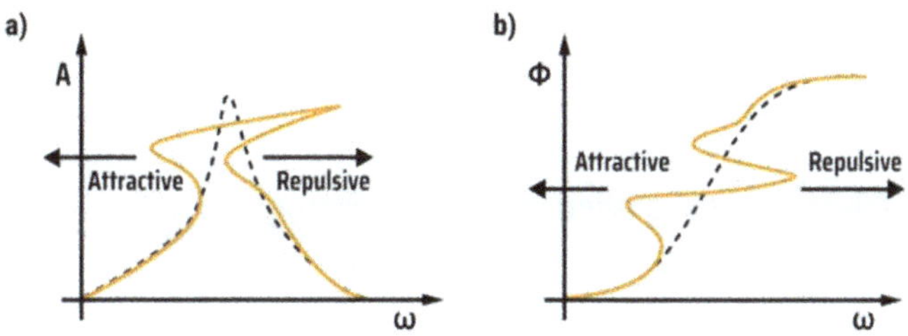

Figure 5.4. Nonlinear case. Illustration of a standard frequency sweep showing the (a) amplitude A and (b) phase ϕ response where the origin of the distortions is shown in terms of whether nonlinear forces are attractive or repulsive. Attractive forces pull the curves to lower frequencies while repulsive forces pull them towards higher frequencies.

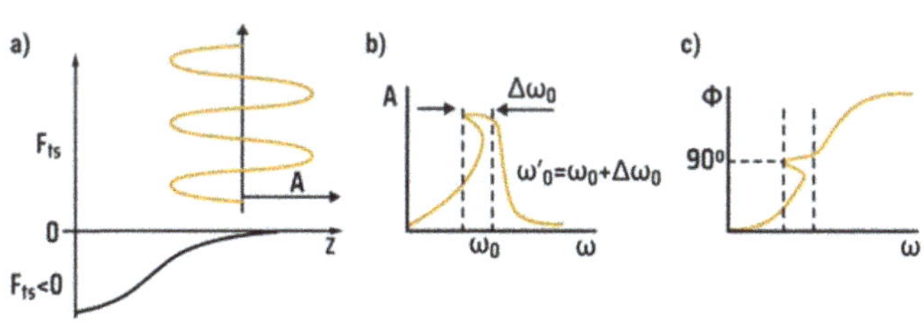

Figure 5.5. Nonlinear case. (a) Illustration of a nonlinear (attractive) force showcasing how a sinusoidal wave, i.e., the motion of the oscillator, would be affected by it. Corresponding illustration of a standard frequency sweep showing the (b) amplitude A and (c) phase ϕ response where the origin of the distortions is shown in terms of nonlinear attractive forces only.

5.3 The linear response

The response of the linear expression is given by the expression for the amplitude A and the phase shift ϕ. The phase shift is the difference in phase between the amplitude response and the drive force $F_0 \cos \omega t$. The position z is parametrized as

$$z(\omega) = A\cos(\omega - \phi)$$

Where the amplitude is given by equation (4.22)

$$A(\omega) = \frac{F_0}{\sqrt{m^2(\omega_0^2 - \omega^2)^2 + b^2\omega^2}}$$

and the phase shift is given by equation (4.22)

$$\tan \phi(\omega) = \frac{b\omega}{m(\omega_0^2 - \omega^2)}$$

The reader should note that the amplitude and phase versus frequency ω curves shown in figures 5.2–5.7 are only sketches and 'real' curves are typically found by numerically solving the equation of motion [7]. This point will be further discussed in the next chapter when discussion the derivation of the virial of interaction V. While the linear response has been discussed analytically in section 5.1, the amplitude and phase response as a function of drive frequency, i.e., ω, had not been shown before. Some notes are relevant in order to later interpret the nonlinear response. First, in AFM the drive frequency can be set by the user. In other systems ω is given by other phenomena which is not controlled in the experiment.

(1) If $\omega < \omega_0$, then $\phi < 90°$,
(2) If $\omega = \omega_0$, then $\phi = 90°$,
(3) If $\omega > \omega_0$, then $\phi > 90°$.

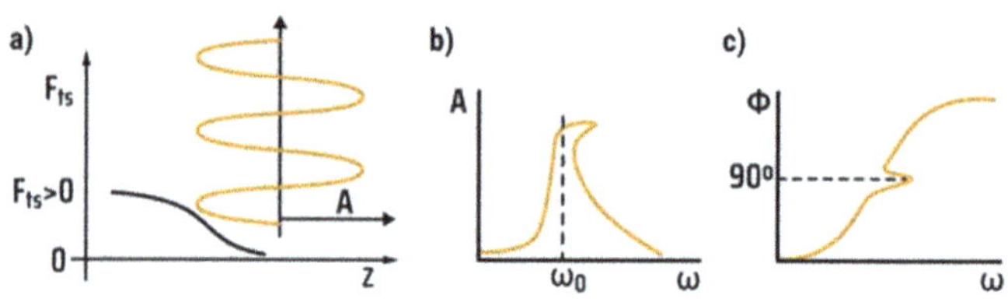

Figure 5.6. Nonlinear case. (a) Illustration of a nonlinear (repulsive) force showcasing how a sinusoidal wave, i.e., the motion of the oscillator, would be affected by it. Corresponding illustration of a standard frequency sweep showing the (b) amplitude A and (c) phase ϕ response where the origin of the distortions is shown in terms of nonlinear repulsive forces only.

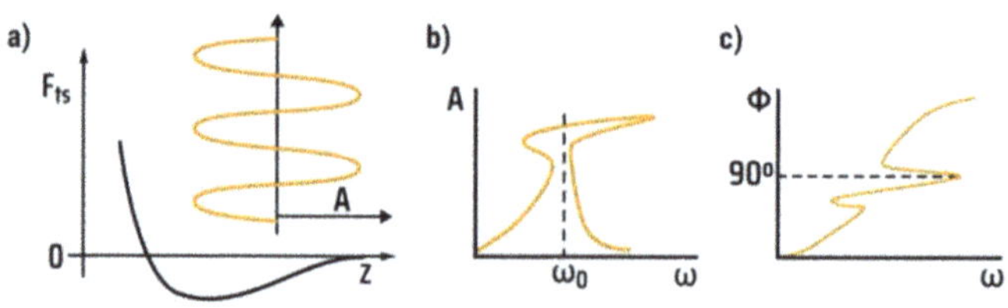

Figure 5.7. Nonlinear case. (a) Illustration of a nonlinear force with attractive and repulsive components showcasing how a sinusoidal wave, i.e., the motion of the oscillator, would be affected by it. Corresponding illustration of a standard frequency sweep showing the (b) amplitude A and (c) phase ϕ response where the origin of the distortions is shown in terms of the presence of nonlinear attractive and repulsive forces.

For the linear equation in equation (5.4) there is no constraint in the above results. This means that the results are true independently of Q, F_0, k, b, ω_0 and m. The same is true for ϕ. Namely, ϕ is determined by ω alone for a given set of parameters Q, F_0, k, b, ω_0 and m. The same holds for the amplitude A, namely, for a given set of parameters Q, F_0, k, b, ω_0 and m, the maximum amplitude occurs at equation (4.27)

$$\omega_r \,|_{A=A_{\max}} = \omega_0\left[1 - \frac{1}{2Q^2}\right]^{1/2}$$

The maximum amplitude $A_{\max}$ is usually also termed A_0 in AFM. The amplitude behaves as follow

(1) If $\omega < \omega_r$, then $A < A_0$ and A monotonically decreases with decreasing ω,
(2) If $\omega = \omega_r$, then $A = A_0$,
(3) If $\omega > \omega_r$, then $A > A_0$ and A monotonically decreases with increasing ω.

The amplitude A further takes on a single value at each ω. We recall that '[I]n mathematics, a function from a set X to a set Y assigns to each element of X exactly one element of Y'. In terms of equations (4.21) and (4.22) for the linear equation this means that each value of ω maps exactly to one value of A and ϕ, respectively. We will see that this is not the case in the nonlinear case.

5.4 The nonlinear response and the coexistence of several stable states of oscillation

Already in 1991 [8] it was reported that a tip vibrating near a surface would lead to bi-stable behaviour. Here, bi-stability means that for a set of parameters, the oscillation will stabilize, i.e., the steady state of oscillation might be reached, at two different values of A [4, 5, 7–9]. In the linear case, for a given ω the response z is

$$z(\omega) = A\cos(\omega t - \phi) \tag{5.8}$$

The expression in equation (5.8) satisfies the definition of mathematical function as each ω maps to a unique z. Namely, the definition '[I]n mathematics, a function from a set X to a set Y assigns to each element of X exactly one element of Y' is satisfied between ω and z in equation (5.8).

The presence of bi-stability, or several oscillation states, means that for a given ω there will be at least two stable solutions. For example, a low L and high H branch for z can be written as follows for a given ω

$$z_L(\omega) = A_L\cos(\omega t - \phi_L)$$

$$z_H(\omega) = A_H\cos(\omega t - \phi_H) \tag{5.9}$$

The above expressions, i.e., equation (5.9), show that the mapping from ω to z does not satisfy the definition of a mathematical function expressed above. The implication is that the coexistence of several oscillation states leads to a mapping of ω to z that cannot be represented as a single function (figure 5.3). It will be later shown that analytical expressions for A in terms of ω are no longer of the form where a single A results for every ω. These expressions are thus consistent with the existence of two stable oscillation states. In figure 5.3 three values of A and ϕ can be observed for some values of ω but, in such cases, the middle one is not physically accessible [7]. Such phenomena was reported by numerically solving the equation of motion instead [7]. In figure 5.3 the amplitude A and phase ϕ are distorted versions of the linear response shown in figure 5.2.

Attractive and repulsive forces are responsible for distorting the amplitude A and phase ϕ versus ω curves to lower and higher values of ω, respectively, as shown schematically in figure 5.4. When only attractive forces distort the curves toward lower frequencies only an oscillation branch, where F_{ts} is not negligible, is accessible a range of values of ω. There might be another oscillation branch for which F_{ts} is negligible at the same range of ω. This other branch coincides with the branch produced by the linear response. When repulsive forces distort the curve, the phenomenon is mirrored toward higher values of ω. When there are attractive and repulsive forces present there are two distortions that contribute to the total distortion and the accessible branches of z are accessible for which F_{ts} is not negligible. These are represented by equation (5.9).

5.5 The nonlinear response (attractive case)

The first type of nonlinearity to discuss is that shown in figure 5.5 as expressed by equation (5.5). Here, the force is attractive as expressed by the negative sign in the equation, i.e., $F_{ts} \propto -\alpha z^{-2}$. If the force varied linearly with z it could be absorbed by the kz term and the solution would reduce to the case discussed in section 3.2. Nevertheless, here the differential equation is nonlinear because one of the terms, i.e., F_{ts}, is expressed in terms of z, the dependent variable, as a power. It is assumed that α is a positive constant (equation (2.5)).

Several points are worth mentioning

(1) The oscillation amplitude A depends on ω (figure 2.3) but also on the equilibrium position. This is because if the force is proportional to $\propto z^{-2}$, the way in which the force affects the amplitude depends on the actual values of z^{-2}. See section 3.2 when discussing distance independent forces to see that linearity implies that only the reference point for which $kz = 0$ varies in these cases. For example, a linear term such as kz produces a differential of force that is independent of the actual value of z

$$F = kz = > dF = \mathrm{kdz} \tag{5.10}$$

The above equation expressed proportionality with dz. The magnitude of the differential of the force depends on k, i.e., a constant only. For the nonlinear case, i.e., $-\alpha z^{-2}$, the differential of F gives

$$F = -\frac{\alpha}{z^2} = > dF = 2\frac{\alpha}{z^3}dz \tag{5.11}$$

Equation (5.11) shows that the differential dF depends on z and on the reference value of z. This means that if a new reference value for which $kz = 0$ is considered, the net force in equation (5.11) will also vary. For this reason the integral over an oscillation cycle for a given A depends on the equilibrium position z_0, that is, on the absolute values of z that the mass covers in its motion. For the linear case the total range covered by the point-mass system is $2A$, i.e., peak to peak. This is clear from the fact that $z = A\cos(\omega t - \phi)$. In the linear case the absolute values of z are irrelevant and only relative values matter since the instantaneous (conservative) net force is always kz as measured from the reference point $kz = 0$ (see section 3.2). For the nonlinear case the range covered is also $2A$ but the instantaneous net (conservative) force depends on the actual values of z. This means that if the equilibrium position is displaced the forces acting on the mass for the range of motion under consideration also change (figure 5.5).

(2) Neither the amplitude curve A, nor the phase curve ϕ, are a mathematical function any longer since the amplitude A no longer takes on a single value at each ω. Rather the curve deforms towards lower values of ω (figures 5.3–5.5).

(3) The phase = 90° for values of $\omega < \omega_0$. This is due to the deformation of the curve. In particular, a new 'natural' frequency of resonance ω_0' can be defined where $\omega_0' = \omega_0 - \Delta\omega_0$. The fact that the force is attractive accounts for the deformation toward lower values of ω as discussed when interpreting figure 5.4.

(4) The shape of the curve, and therefore the value of ω_0' depends on the equilibrium position. Because of what has been said in point 1, the deformation of the curve also depends on the actual values of force and on the equilibrium position.

5.6 The nonlinear response (repulsive case)

The second type of nonlinearity to discuss is that shown in figure 5.6 and expressed by equation (5.6). The force is repulsive, i.e., always pushes the mass toward higher values of z. Again, the differential equation is nonlinear because z, the dependent variable, is expressed as a power, i.e., $\propto z^{\frac{3}{2}}$, in one of its terms. Here (equation (5.6) β is a constant. Again, if the force varied linearly with z it could be absorbed by the kz term.

Several points are worth mentioning for the repulsive case

(1) The oscillation amplitude A depends on ω (figure 5.6) but also on the equilibrium position. This is because if the force is proportional to $\propto z^{-2}$, the way in which the force affects the amplitude depends on the actual values of z and the equilibrium position. See section 3.2 when discussing distance independent forces. The interpretation is similar to that give in point one above for the attractive case.

(2) The amplitude A and phase ϕ curves are not a mathematical function any longer (figures 5.3, 5.4 and 5.6) since the amplitude A no longer takes on a

single value at each ω. Rather the curve deforms towards higher values of ω (figure 5.6).

(3) The phase = 90° for values of $\omega > \omega_0$. This is due to the deformation of the curve. In particular, a new 'natural' frequency of resonance ω_0' can be defined where $\omega_0' = \omega_0 + \Delta\omega_0$.

(4) The shape of the curve, and therefore the value of ω_0' depends on the equilibrium position. The deformation of the curve also depends on the force and the equilibrium position as in the attractive nonlinear case.

5.7 The nonlinear response (attractive/repulsive case)

This case is expressed by equation (5.7). The behaviour of the amplitude A and the phase ϕ are shown in figure 5.7. The force has an attractive and a repulsive term. The force combines the expressions in equations (5.5) and (5.6) as illustrated in figure 5.7.

5.8 Summary

The above figures 5.3–5.7 and the respective nonlinear equations of motion, i.e., equations (5.5)–(5.7), have been descriptively discussed only since the nonlinear terms must be well parametrized in order to avoid divergence and to determine their value as a function of z to avoid other similar problems. Typically, the parametrization or determination will be established from the phenomena being modelled. Note for example that equations (5.5)–(5.7) are not well defined for $z = 0$ because there is a term that divides by zero. In AFM the physical interpretation to avoid this problem is that there cannot be matter interpenetration. That is, two bodies cannot occupy the same space. The distance from the tip to the sample will determine the actual values of z. Otherwise the discussion above is general.

Equations (5.5)–(5.7) will be parametrized below in order to better analyse such phenomenon and understand the models and the physical system in AFM.

Finally, it is worth remembering that the physical representation of the variables in the equations of motion (equations (5.5)–(5.7)) do not have to be position or time. The implication is that any phenomena that can be represented by a model mathematically equivalent to the above can also be analysed via the discussion provided here and extended below.

References

[1] Westphal W H 1968 Mechanics of point masses and rigid bodies *A Short Textbook of Physics: Not Involving the Use of Higher Mathematics* ed W H Westphal (Berlin: Springer) pp 6–57

[2] Wikipedia Nonlinear system https://en.wikipedia.org/wiki/Nonlinear_system

[3] Colton J S 2020 Lorentz Oscillator Model *Physics Course at Brigham Young University* 442–71

[4] Levi A F J 2016 The Lorentz Oscillator Model In *Essential Classical Mechanics for Device Physics* (San Rafael, CA: Morgan & Claypool Publishers) pp 5-1–5-21

[5] Almog I F, Bradley M S and Bulović V 2022 The Lorentz Oscillator and Its Applications https://docplayer.net/48207190-The-lorentz-oscillator-and-its-applications-described-by-i-f-almog-m-s-bradley-and-v-bulovic.html
[6] Raman A 2014 Atomic force microscopy https://nanohub.org/resources/520
[7] Paulo Á S and García R 2002 Unifying theory of tapping-mode atomic-force microscopy *Phys. Rev.* B **66** 041406
[8] Gleyzes P, Kuo P K and Boccara A C 1991 Bistable behavior of a vibrating tip near a solid surface *Appl. Phys. Lett.* **58** 2989–91
[9] Marth M, Maier D, Honerkamp J, Brandsch R and Bar G 1999 A Unifying View on Some Experimental Effects in Tapping-Mode Atomic Force Microscopy *J. Appl. Phys.* **85** 7030–6

IOP Publishing

Oscillations

Theory and applications in AFM
Tuza Adeyemi Olukan, Sergio Santos, Lamiaa Sami Elsherbiny and Matteo Chiesa

Chapter 6

The cantilever and tip–sample system in AFM

6.1 Forces

The discussion above (chapter 5) and the interpretation of figures 5.3–5.7 is general but the nonlinear terms in equations (5.5)–(5.7) have not been well defined yet. The definition or determination of these terms depends on the experimental set-up or, theoretically, on the parameters of the model being considered.

Here we proceed to constrain the parameters of the nonlinear terms in order to understand more clearly how the amplitude and phase response vary as a function of ω. For this purpose we use a typical model for the tip–sample force employed in AFM. In AFM the distance d is measured from the surface of the sample to the tip. The schematic of the geometry of the cantilever–tip–sample system is shown in figure 6.1. A main constraint worth mentioning is that $d = z_c + z$.

6.1.1 The attractive force

The attractive force is typically modelled by invoking the van der Waals force considering an approach first developed by Hamaker [1, 2]. This is a force of attraction since it tends to pull surfaces towards each other (figure 5.7). The surface here provides the constraint that determines a zero distance d. In this model, the tip is modelled a sphere of radius R where R parametrizes the geometry of the tip. The force is proportional to H (Hamaker) where H depends on the properties of the tip and the sample. Then the attractive force is

$$d = z_c + z$$

$$\alpha = \frac{RH}{6}$$

$$F_{ts} = -\frac{\alpha}{d^2} \quad d > a_0$$

doi:10.1088/978-0-7503-5809-5ch6

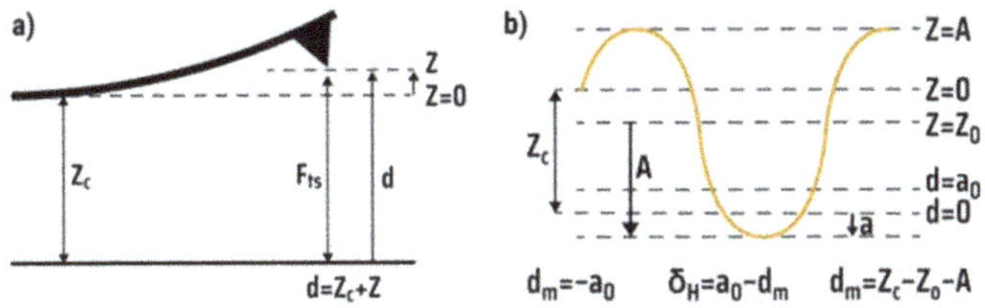

Figure 6.1. (a) Schematic of an AFM cantilever from which geometrical constraints can be derived. (b) Illustration of the tip' motion showcasing the geometric parametrization.

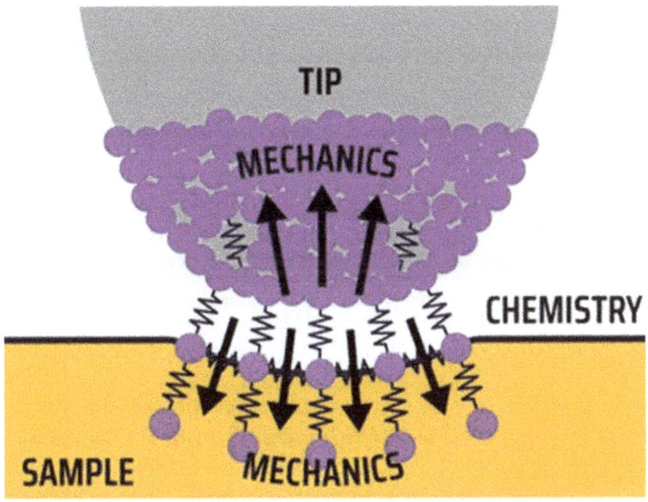

Figure 6.2. Illustration showcasing the interaction between an AFM tip and a surface where attractive and repulsive forces result from chemical and mechanical interactions. See a detailed discussion in the literature [3]. Reprinted with permission from [3]. Copyright (2013) American Chemical Society.

$$F_{ts} = -\frac{\alpha}{a_0^2} = F_{AD} \quad d \leqslant a_0 \tag{6.1}$$

Equation (6.1) has a parameter a_0, which physically represents an intermolecular distance that impedes matter interpretation [4], i.e., the tip's and the sample's atoms cannot occupy the same distance (figure 5.7). At $d = a_0$ we have minima for F_{ts}, or the force of adhesion F_{AD}. Beyond a_0 the surfaces cannot physically come any closer so mechanical deformation occurs. An illustration of such phenomenon is shown in figure 6.2. This implies that negative values of d are possible but physically these correspond to sample deformation δ [3]. The force profile (or nonlinear term) resulting from equation (6.1) is illustrated in figure 6.3.

The parameter a_0 in equation (6.1) solves the problem of divergence since the nonlinear term in equation (5.5) never divides by zero because d is never zero.

If there was no other force and if the tip oscillated only at distances $d < a_0$, the attractive force would only shift the equilibrium position and the equation would reduce to a linear equation of motion (see section 3.2).

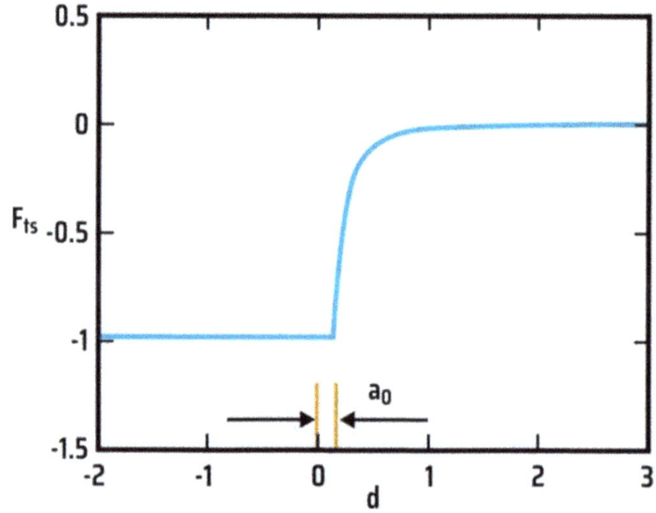

Figure 6.3. Example of an attractive force F_{ts} shown as a function of distance d and representing the force in equation (6.1). a_0 is an intermolecular distance that ensures that d is never zero, i.e., two bodies never occupy the same space.

6.1.2 The repulsive force

From the schematic in figure 5.6, the general equation for the tip–sample instantaneous distance d (ignoring higher harmonics and modes) can be written as

$$z = z_0 + A\cos(\omega t - \phi)$$

$$d = z_c + z_0 + A\cos(\omega t - \phi) \tag{6.2}$$

where z_0 is the mean deflection and negative if the mean force $\langle F_{ts} \rangle$ is negative and positive otherwise. From the schematic in figure 5.7 the minimum distance of approach is given when $z = -A$, then

$$d_m = z_c + z_0 - A \tag{6.3}$$

Typically the mean deflection z_0 is much smaller than A so that

$$d_m \approx z_c - A \tag{6.4}$$

The cantilever-sample separation z_c can be written in terms of physically meaningful parameters from equations (6.3) and (6.4)

$$z_c = d_m - z_0 + A \tag{6.5}$$

Assume now that $z_c + a_0 < A + |z_0|$ where $z = A\cos(\omega t - \phi)$. Then there are values of d during an oscillation for which $d < a_0$. At this point mechanical contact occurs, that is, there is mechanical deformation $\delta > 0$. The deformation can be written as

$$\delta = a_0 - d \quad d \leqslant a_0 \tag{6.6}$$

The maximum of deformation δ_M can be written as

$$\delta_M = a_0 - d_m \tag{6.7}$$

In terms of observables A and z_0,

$$d = d_m + A + A\cos(\omega t - \phi) \tag{6.8}$$

$$\delta = a_0 - d \quad d \leqslant a_0 \tag{6.9}$$

The repulsive force in equations (5.6) and (5.7) is typically expressed in AFM as follows [1, 5–7]

$$d = z_c + z$$

$$\beta = \frac{4}{3}E'\sqrt{R}$$

$$\delta = a_0 - d \quad d \leqslant a_0$$

$$\beta\delta^{3/2} \tag{6.10}$$

where E' is the reduced elastic modulus of tip and sample pair and R is the effective radius of the tip. With the above set of expression thc rcpulsivc tcrm is parametrized and physically determined in AFM. Namely, the term is non-zero and positive only where there is sample deformation. $\delta \geqslant 0$ (figures 6.1 and 6.2).

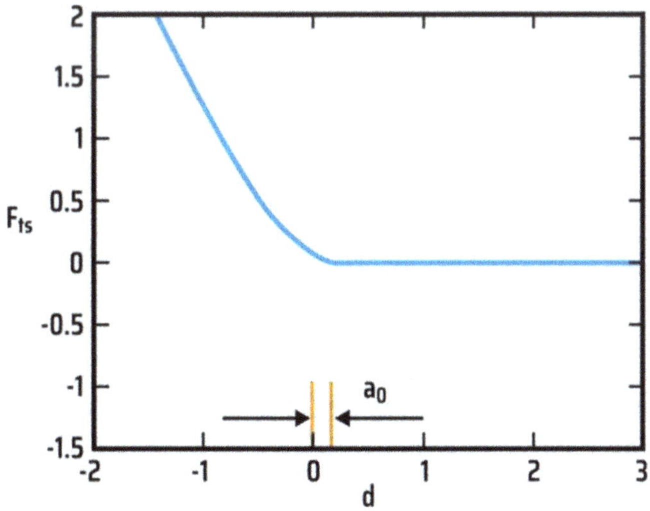

Figure 6.4. Example of a repulsive force F_{ts} shown as a function of distance d and representing the force in equation (6.10). a_0 is an intermolecular distance that ensures that d is never zero, i.e., two bodies never occupy the same space.

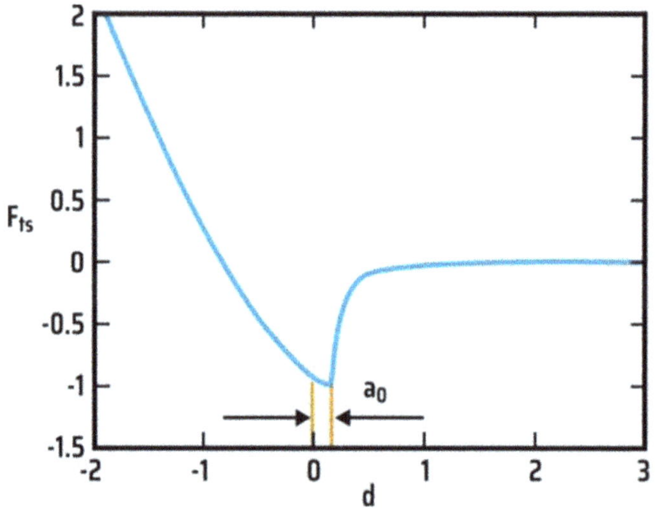

Figure 6.5. Example of attractive and repulsive forces F_{ts} shown as a function of distance d and representing the force in equation (6.11). a_0 is an intermolecular distance that ensures that d is never zero, i.e., two bodies never occupy the same space.

6.1.3 The tip–sample interaction: attractive–repulsive components

The expressions in equations (6.1) and (6.10) can be added to give a force profile (figure 6.3) identified with the Dejarguin–Muller–Toporov (DMT) theory of contact mechanics [8, 9] where

$$F_{ts} = -\frac{RH}{6d^2} \quad d > a_0$$

$$F_{ts} = -F_{AD} + \frac{4}{3}E'\sqrt{R}\,\delta^{3/2} \quad d \leqslant a_0 \tag{6.11}$$

The profile shown in figure 6.5 displays both attractive and repulsive nonlinearities. This agrees with the claim that, in AFM, the amplitude and phase curves expressed in terms of ω may present distortions as those shown in figures 5.3–5.7. More complex models are sometimes discussed in the literature. In particular, dissipative forces may also distort the curves. The analytical expressions leading to such distortions in terms of conservative and dissipative contributions will be the object of the next chapters.

6.2 Conservative and dissipative terms: the virial and energy dissipation

Here the theory of nonlinear interactions is developed by solving the equation of motion in equation (5.3). The equation is reproduced here to examine the terms

$$m\frac{d^2z}{dt^2} + \frac{m\omega_0}{Q}\frac{dz}{dt} + kz = F_{ts} + F_0 \cos \omega t \tag{6.12}$$

6.2.1 Energy dissipation

In 1998 Tamayo and García [10, 11] and Cleveland, Anczykowski, Schmid, and Elings [12, 13] independently developed a formalism based on equation (6.12) to compute an expression for the energy dissipated in the tip–sample interaction. This energy is dissipated because of the dissipative component of F_{ts} and is to be distinguished from the dissipation due to the linear term $\frac{m\omega_0}{Q}\frac{dz}{dt}$. Cleveland *et al* focused on the average power dissipated per cycle while Tamayo and García focused on the average energy dissipated per cycle. Physically, it is arguably more relevant to speak of average power, that is, of energy per second, but energy dissipation is meaningful when one speaks of physical processes. For example, atomic phenomena is typically considered in eV units and the energy dissipation per cycle in dynamic AFM is also expressed in these units, i.e. $\langle E_{dis}\rangle \sim 1\text{–}100$ eV[29]. In any case moving from energy to power is straightforward

$$\langle P_{dis}\rangle_{cycle} = \frac{\langle E_{dis}\rangle_{cycle}}{T} \tag{6.13}$$

Here we will simplify the notation by writing

$$\langle E_{dis}\rangle_{cycle} \equiv E_{dis} \tag{6.14}$$

The energy dissipated per cycle can be obtained directly by considering that this is the work done by the force F_{ts} per cycle. The period T determines the length of a cycle in terms of time. Then

$$E_{dis} = -\int_0^T F_{ts}\dot{z}dt \tag{6.15}$$

where it is assumed that (compare with equation (4.24)) the response is harmonic

$$z = z_0 + A\cos(\omega t - \phi)$$

$$z \approx A\cos(\omega t - \phi) \tag{6.16}$$

Derivatives over time produce

$$\dot{z} = -\omega A\sin(\omega t - \phi) \tag{6.17}$$

$$\ddot{z} = -\omega^2 A\cos(\omega t - \phi) \tag{6.18}$$

Note that while $\varphi < 0$, we have taken the convention of assigning the sign already in equation (6.16). Then the angle, i.e., the phase shift between the drive and the response will be taken as an absolute value. Equation (6.15) can be derived from unknowns by rearranging equation (6.12).

$$m\frac{d^2z}{dt^2} + \frac{m\omega_0}{Q}\frac{dz}{dt} + kz = F_0\cos\omega t + F_{ts}$$

$$-F_{ts} = F_0 \cos \omega t - m\frac{d^2z}{dt^2} - \frac{m\omega_0}{Q}\frac{dz}{dt} - kz$$

$$E_{dis} = -\int_0^T F_{ts}\dot{z}dt$$

$$E_{dis} = \int_0^T \left[F_0\cos \omega t - m\frac{d^2\dot{z}}{dt^2} - \frac{m\omega_0}{Q}\frac{dz}{dt} - kz\right]\dot{z}dt \tag{6.19}$$

The term z only appears as kz. The other terms only contain $\dot{z}$ or $\ddot{z}$. Note that all the terms on the right are identical to the terms of the linear theory. Compare, for example, equation (5.2) with equation (6.20) below

$$m\frac{d^2z}{dt^2} + \frac{m\omega_0}{Q}\frac{dz}{dt} + kz = F_0 \cos \omega t + F_{ts}$$

$$0 = F_0 \cos \omega t - \left[m\frac{d^2z}{dt^2} + \frac{m\omega_0}{Q}\frac{dz}{dt} + kz\right] F_{ts} = 0 \tag{6.20}$$

Then, E_{dis} is found by solving the integrals of the four linear terms (the knowns) as follows:

$$E_{dis} = -\int_0^T F_{ts}\dot{z}dt$$

$$E_{dis} = \int_0^T \left[F_0\cos \omega t - m\frac{d^2\dot{z}}{dt^2} - \frac{m\omega_0}{Q}\frac{dz}{dt} - kz\right]\dot{z}dt$$

$$E_{dis} = I_1 + I_2 + I_3 + I_4 \tag{6.21}$$

Note that if the integrals of the four terms on the right cancel out, $E_{dis} = 0$. Each of the integrals can be easily solved by considering orthogonality

$$I_1 = \int_0^T F_0 \cos \omega t\dot{z}dt$$

$$I_1 = -\omega F_0 A \int_0^T [\sin(\omega t)\cos\phi - \cos(\omega t)\sin\phi]\cos\omega t dt$$

$$I_1 = \omega F_0 A \int_0^T \cos^2(\omega t)\sin\phi dt$$

$$I_1 = \omega F_0 A \sin\phi \frac{T}{2}$$

$$I_1 = \pi F_0 A \sin\phi \tag{6.22}$$

$$I_2 = -\int_0^T m\frac{d^2z}{dt^2}\dot{z}dt$$

$$I_2 = \omega^3 A^2 m \int_0^T \cos(\omega t - \phi)\sin(\omega t - \phi)dt$$

$$I_2 = \omega^3 A^2 m \int_0^T [-\cos^2(\omega t)\cos\phi \sin\phi + \sin^2(\omega t)\cos\phi \sin\phi]dt$$

$$I_2 = \omega^3 A^2 m \cos\phi \sin\phi \int_0^T [\sin^2(\omega t) - \cos^2(\omega t)]dt$$

$$I_2 = \omega^3 A^2 m \cos\phi \sin\phi \left[\frac{T}{2} - \frac{T}{2}\right]$$

$$I_2 = 0 \tag{6.23}$$

$$I_3 = -\int_0^T \frac{m\omega_0}{Q}\dot{z}^2 dt$$

$$I_3 = -\frac{m\omega_0}{Q}A^2\omega^2 \int_0^T \sin^2(\omega t - \phi)dt$$

$$I_3 = -\frac{m\omega_0}{Q}A^2\omega^2\left[\frac{T}{2}[\sin^2\phi + \cos^2\phi]\right]$$

$$I_3 = -\frac{m\omega_0}{Q}\omega\pi A^2 \tag{6.24}$$

$$I_4 = -k\int_0^T z\dot{z}dt$$

$$I_4 = -k\int_0^T [z_0 + A\cos(\omega t - \phi)]\dot{z}dt$$

$$I_4 = kA^2\omega \int_0^T \cos(\omega t - \phi)\sin(\omega t - \phi)dt - kz_0\int_0^T \dot{z}dt$$

$$I_4 = 0 \tag{6.25}$$

We note from the above result that z_0 does not add to E_{dis}. Finally, combining equations (6.21)–(6.25)

$$E_{dis} = \pi F_0 A \sin\phi - \frac{m\omega_0}{Q}\omega\pi A^2 \tag{6.26}$$

where the term containing the sign is always positive or zero since $0° \leqslant \varphi \leqslant 180°$. The last term is always negative because there is a minus sign and all the terms involved are positive. Note that F_0 depends on ω amongst others and it can be experimentally found by considering the expression for A from the linear results (equation (4.21)). The above expression already gives us a condition for $E_{dis} = 0$

$$0 = \pi F_0 A \sin\phi - \frac{m\omega_0}{Q}\omega\pi A^2$$

$$\sin\phi = \omega\frac{m\omega_0}{Q}\frac{A}{F_0}$$

$$\sin\phi = \omega b\frac{A}{F_0} \tag{6.27}$$

When $\omega = \omega_0$

$$\sin\phi = \frac{m\omega_0^2}{Q}\frac{A}{F_0}$$

$$\sin\phi = \frac{k}{Q}\frac{A}{F_0}$$

$$\sin\phi = \frac{A}{A_0}$$

$$\phi = \sin^{-1}\left(\frac{A}{A_0}\right) \tag{6.28}$$

The only non-zero terms are the one related to viscosity, i.e., Q, and the one derived from the drive. This is because in the steady state the kz term and the inertia terms remain constant and neither add nor subtract from the energy delivered by the driving force. In the steady state the drive is the only term delivering energy to the system and the only terms dissipating are the linear viscosity and the nonlinear part of F_{ts} provided there are dissipative forces introduced by F_{ts}. This is intuitive and equation (6.26) simply confirms this.

The above expression provides two branches for $E_{dis} = 0$ since $\sin\varphi$ is symmetric around $\varphi = 90°$. The phase shift is thus prescribed from A and A_0 alone if $E_{dis} = 0$. This is shown in figure 6.6. In particular, in dynamic AFM A_0 is the free amplitude obtained when there is no tip–sample interaction. In AM AFM A can be set as a target by the user. The amplitude A is reduced when there is interaction.

The following equation also applies because of the phenomenon being analysed cannot act as a drive, i.e., can never make the amplitude larger than the amplitude when $F_{ts} = 0$. Then,

$$0 \leqslant \frac{A}{A_0} \leqslant 1 \tag{6.29}$$

In 2012 Gadelrab *et al* [14] discussed that only values of phase shift φ lying within the function in figure 6.6 are physically possible. This is because values of φ lying

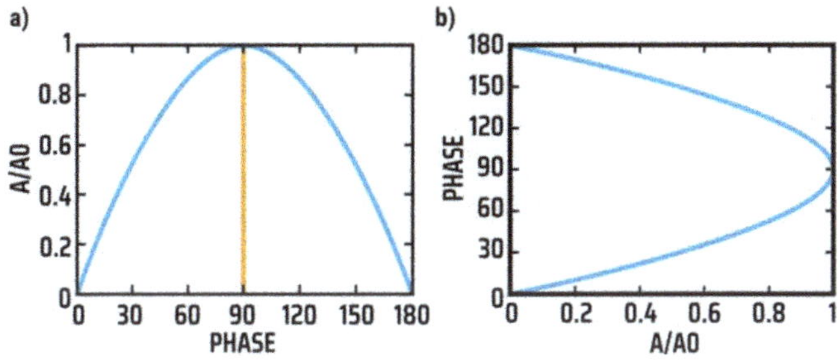

Figure 6.6. Nonlinear response. (a) Graph displaying the behaviour of the amplitude A normalized in terms of the unperturbed or free amplitude A_0 as a function of phase ϕ. The graph shows the behaviour when only conservative nonlinear terms are present. Two branches can be observed. In AFM these are called attractive and regime regimes, but they can be identified with the available states, H and L, in figure 2.3(b). The same graph plotted in terms of A/A_0 in the x-axis. See the literature for more details [14].

above the curve would imply that the system is generating energy. This is against energy conservation and thus impossible. For example, since the forces in equation (6.11) are conservative, for a given amplitude A and a given drive force producing A_0 the constraints in phase φ in equations (6.28) and (6.29) must apply. This means that any contrast in phase φ images when keeping A, A_0 and ω constant must be due to dissipative interactions. The constraint can be written as follows:

There will be dissipation as indirectly observed from phase shifts φ whenever

$$\frac{A}{A_0} < \sin \phi \tag{6.30}$$

This is interesting since it provides a means to interpret phase contrast in AM AFM and also in frequency modulation FM AFM. In AM AFM A is constant while imaging while in FM AFM it might vary. From the above discussion it follows that

(1) In AM AFM, heterogeneity in dissipative interactions will be observed as phase contrast maps.
(2) Whenever $A/A_0 < \sin \varphi$, it can be concluded that there is dissipation in the tip–sample interaction. Mechanisms of dissipation can thus be investigated.
(3) Furthermore, for a given set of driving parameters A, A_0 and ω, a difference of phase shift can be defined as $\Delta\varphi$ where

$$\Delta\phi = \phi - \sin^{-1}\left(\frac{A}{A_0}\right) \tag{6.31}$$

The larger the above in a phase contrast map, the more dissipation. This provides an indirect method to visually interpret energy dissipation in phase contrast images. For example, in figure 6.7 a phase contrast image obtained in AM AFM was investigated [14] where the system was a carbon nanotube (CNT) on a quartz surface. By plotting $\Delta\varphi$ instead of φ the data clearly shows that more energy is dissipated on the CNT than on the quartz surface.

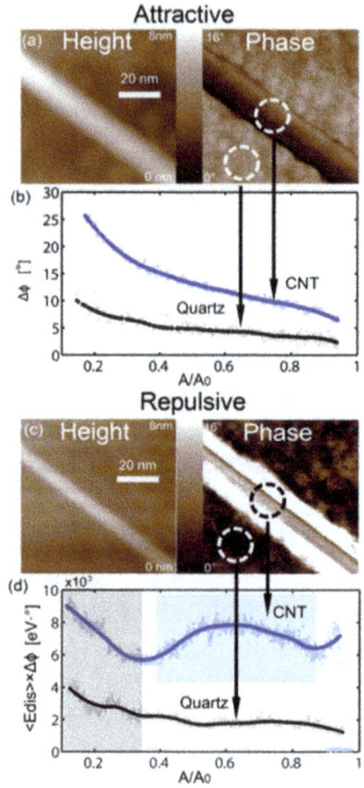

Figure 6.7. (a) Height and phase contrast for a CNT on a quartz surface obtained in the attractive regime. (b) The phase difference $\Delta\phi$ from equation (6.31) is plotted as a function of A/A_0. (c) Height and phase contrast for the same CNT and quartz system obtained in the repulsive regime. (d) The product $E_{dis}\,\Delta\varphi$ is shown to lead to distinctive maxima at intermediate A/A_0 values for viscous and hysteretic dissipation. See the literature for details [14]. Reproduced from [14].

Finally, the energy dissipation expression in equation (6.26) can be written more compactly for $\omega = \omega_0$ as

$$E_{dis}(d) = \frac{\pi k A_0 A(d)}{Q}\left[\sin\,(\Phi(d)) - \frac{A(d)}{A_0}\right] \tag{6.32}$$

where the expression emphasizes that both A and φ are a function of d or d_m (see schematic in figure 6.1). The energy dissipation expression can also be interpreted as the F_{ts} term doing work on the mass-spring system does reducing its energy. From this interpretation it follows that the work done should be written in terms of a negative sign

$$E_{dis} = \int_0^T F_{ts}\dot{z}dt \tag{6.33}$$

Here E_{dis} would be negative, but the interpretation clarifies that it is the work done by the force on the system, i.e., the work done would be less than zero because F_{ts} can only remove energy from the system.

6.2.2 The virial of the interaction

In 2001 San Paulo and García exploited the virial theorem [7] to find a transfer function from equation (5.3). The virial theorem was introduced for the first time in 1870 by Rudolf Clausius [15] but San Paulo and García referred to a classical book in classical mechanics [16]. The theorem 'provides a general equation that relates the average over time of the total kinetic energy of a stable system of discrete particles, bound by potential forces, with that of the total potential energy of the system' [17]. Mathematically and in relation to equation (5.3), San Paulo and García used the fact that 'the time averaged kinetic energy of the tip is equal to its virial' [7]. Then

$$\langle KE\rangle = \frac{1}{2}m\langle \dot{z}^2\rangle = -\frac{1}{2}\langle F_{ts}z\rangle$$

$$V = \langle F_{ts}z\rangle = \frac{1}{T}\int_0^T F_{ts}z dt \tag{6.34}$$

Here V is the virial of F_{ts}. The average kinetic energy $\langle KE\rangle$ was already computed equation (6.26) in section 3.1, and the method to compute it is equivalent to that obtained when considering the linear system

$$\langle KE\rangle = \frac{1}{T}\int_0^t \frac{1}{2}m\dot{z}^2 dt$$

$$\langle KE\rangle = \frac{1}{2}m\langle \dot{z}^2\rangle \tag{6.35}$$

The method to compute the virial if the same as that used in the previous section. Namely, from the equation of motion equation (5.3)

$$m\frac{d^2z}{dt^2} + \frac{m\omega_0}{Q}\frac{dz}{dt} + kz = F_0\cos\omega t + F_{ts}$$

$$0 = -F_0\cos\omega t + \left[m\frac{d^2z}{dt^2} + \frac{m\omega_0}{Q}\frac{dz}{dt} + kz\right]F_{ts} = 0 \tag{6.36}$$

Then, V is found by solving the integrals of the four linear terms as follows

$$V = \langle F_{ts}z\rangle = \frac{1}{T}\int_0^T \left[-F_0\cos\omega t + m\frac{d^2\dot{z}}{dt^2} + \frac{m\omega_0}{Q}\frac{dz}{dt} + kz\right]zdt$$

$$V = I_1 + I_2 + I_3 + I_4 \tag{6.37}$$

Note that if the integrals of the four terms on the right cancel out $V = 0$. Each of the integrals can be easily solved by considering orthogonality. The term z_0 will only be considered in the integral that contains the kz terms since it can be easily shown from orthogonality that the product of a constant, i.e., z_0, and a sine or cosine integrated over a cycle is 0.

$$I_1 = -\frac{1}{T}\int_0^T F_0 \cos \omega t z dt$$

$$I_1 = -\frac{1}{T}\int_0^T F_0 \cos \omega t[z_0 + A\cos(\omega t - \phi)]dt$$

$$I_1 = -\frac{1}{T}z_0\int_0^T F_0 \cos \omega t dt - \frac{1}{T}z_0\int_0^T F_0 \cos \omega t A\cos(\omega t - \phi)dt$$

$$I_1 = -F_0 A\frac{1}{T}\int_0^T \cos(\omega t - \phi)\cos \omega t dt$$

$$I_1 = -F_0 A\frac{1}{T}\int_0^T [\cos(\omega t)\cos\phi + \sin(\omega t)\sin\phi]\cos \omega t dt$$

$$I_1 = -F_0 A\frac{1}{T}\int_0^T \cos^2(\omega t)\cos\phi dt$$

$$I_1 = -F_0 A\frac{1}{T}\cos\phi\frac{T}{2}$$

$$I_1 = -\frac{1}{2}F_0 A\cos\phi \tag{6.38}$$

$$I_2 = \frac{1}{T}\int_0^T m\frac{d^2z}{dt^2}z dt$$

$$I_2 = \frac{1}{T}\int_0^T m\frac{d^2z}{dt^2}[z_0 + A\cos(\omega t - \phi)]dt$$

$$I_2 = -\frac{1}{T}\omega^2 A m z_0\int_0^T \cos(\omega t - \phi)dt - \frac{1}{T}\omega^2 A^2 m\int_0^T \cos(\omega t - \phi)\cos(\omega t - \phi)dt$$

$$I_2 = -\frac{1}{T}\omega^2 A^2 m\int_0^T \cos(\omega t - \phi)\cos(\omega t - \phi)dt$$

$$I_2 = -\frac{1}{T}\omega^2 A^2 m\int_0^T [\cos(\omega t)\cos\phi + \sin(\omega t)\sin\phi]^2\, dt$$

$$I_2 = -\frac{1}{T}\omega^2 A^2 m\int_0^T [\cos^2(\omega t)\cos^2\phi + \sin^2(\omega t)\sin^2\phi]^2\, dt$$

$$I_2 = -\frac{1}{T}\omega^2 A^2 m\left[\frac{T}{2}\cos^2\phi + \frac{T}{2}\sin^2\phi\right]$$

$$I_2 = -\frac{1}{2}\omega^2 A^2 m \tag{6.39}$$

$$I_3 = \frac{1}{T}\int_0^T \frac{m\omega_0}{Q}\dot{z}zdt$$

$$I_3 = \frac{1}{T}\int_0^T \frac{m\omega_0}{Q}\dot{z}[z_0 + A\cos\dot{}(\omega t - \phi)]dt$$

$$I_3 = -\frac{1}{T}\frac{m\omega_0}{Q}Az_0\omega\int_0^T \sin(\omega t - \phi)dt - \frac{1}{T}\frac{m\omega_0}{Q}\omega A^2\int_0^T \sin(\omega t - \phi)\cos(\omega t - \phi)dt$$

$$I_3 = -\frac{1}{T}\frac{m\omega_0}{Q}\omega A^2\int_0^T \sin(\omega t - \phi)\cos(\omega t - \phi)dt$$

$$I_3 = \frac{1}{T}\frac{m\omega_0}{Q}\omega A^2\cos\phi\sin\phi\left[\frac{T}{2} - \frac{T}{2}\right] \tag{6.40}$$

$$I_4 = \frac{1}{T}k\int_0^T z^2dt$$

$$I_4 = \frac{1}{T}k\int_0^T [z_0 + A\cos(\omega t - \phi)]^2\, dt$$

$$I_4 = \frac{1}{T}k\int_0^T [z_0^2 + 2z_0A(\omega t - \phi) + A^2\cos^2(\omega t - \phi)]dt$$

$$I_4 = \frac{1}{T}k\int_0^T z_0^2 \quad dt + \frac{1}{T}k\int_0^T A^2\cos^2(\omega t - \phi)dt$$

$$I_4 = \mathrm{k}z_0^2 + \frac{1}{T}kA^2\int_0^T \cos^2(\omega t - \phi)dt$$

$$I_4 = \frac{1}{T}kA^2\left[\frac{T}{2}[\sin^2\phi + \cos^2\phi]\right]$$

$$I_4 = kz_0^2 + \frac{1}{2}kA^2 \tag{6.41}$$

We note from the above result that z_0 does not add to the dynamic part of V. This can be shown as follows

$$V = \frac{1}{T}\int_0^T F_{ts}zdt$$

$$V = \frac{1}{T}\int_0^T F_{ts}[z_0 + A\cos(\omega t - \phi)]dt$$

$$V = \frac{1}{T}\left[z_0 \int_0^T F_{ts}dt + \int_0^T F_{ts}A\cos(\omega t - \phi)dt\right]$$

$$V = z_0\frac{1}{T}\int_0^T F_{ts}dt + \frac{1}{T}\int_0^T F_{ts}A\cos(\omega t - \phi)dt$$

$$V = z_0\langle F_{ts}\rangle + \frac{1}{T}\int_0^T F_{ts}A\cos(\omega t - \phi)dt \qquad (6.42)$$

But $\langle F_{ts}\rangle = kz_0$, as the reader can confirm by averaging all the terms of equation (5.3) (the equation of motion) over a cycle. Then

$$V = kz_0^2 + \frac{1}{T}\int_0^T F_{ts}A\cos(\omega t - \phi)dt \qquad (6.43)$$

Combining equations (6.38)–(6.41) with equation (6.43)

$$kz_0^2 + \frac{1}{T}\int_0^T F_{ts}A\cos(\omega t - \phi)dt = -\frac{1}{2}F_0A\cos\phi - \frac{1}{2}\omega^2A^2m + kz_0^2 + \frac{1}{2}kA^2 \qquad (6.44)$$

The terms multiplying z_0 cancel out. Thus

$$\frac{1}{T}\int_0^T F_{ts}A\cos(\omega t - \phi)dt = -\frac{1}{2}F_0A\cos\phi + \frac{1}{2}A^2[k - \omega^2m] \qquad (6.45)$$

Typically it is only the term on the left-hand side of equation (6.45) that is termed V for virial in AFM. That is, the mean deflection z_0 does and the respective average force do not count as virial. Then

$$V = \frac{1}{T}\int_0^T F_{ts}A\cos(\omega t - \phi)dt$$

$$V = -\frac{1}{2}F_0A\cos\phi + \frac{1}{2}A^2[k - \omega^2m] \qquad (6.46)$$

When $\omega = \omega_0$

$$V = -\frac{1}{2}F_0A\cos\phi + \frac{1}{2}A^2[k - \omega_0^2m] \text{ any arbitrary } \omega$$

$$V = -\frac{1}{2}F_0A\cos\phi \quad \omega = \omega_0 \qquad (6.47)$$

Already in 2001 when San Paulo and García [7] described the relationship between the virial V and the cosine of the phase shift φ recognized the relationship between this relation (equation (6.47)) and the expression for the frequency shift in FM AFM. In particular, in 1997 Giessibl [18] showed for the first time that the natural frequency ω_0 of a weakly perturbed oscillator shifts according to

$$\frac{\Delta f_0}{f_0} \approx -\frac{1}{kA^2}\langle F_{ts}z\rangle$$

$$\frac{\Delta\omega_0}{\omega_0} \approx -\frac{1}{kA^2}\langle F_{ts}z\rangle$$

$$\frac{\Delta\omega_0}{\omega_0} \approx -\frac{1}{kA^2}V \tag{6.48}$$

The derivation of equation (6.48) exploits the Hamilton–Jacobi formalism, and the reader can refer to the paper by Giessibl to find the details. The expression in equation (6.48) is in agreement with results also reported by San Paulo and García in 2002 in a paper [5] that discussed the distortion of the amplitude A versus ω curves. The authors showed the shape of the curves resulting from numerical integration of the equation of motion. The shapes of the curves produced by numerical integration are similar to the ones illustrated in figures 5.2–5.7. Several points are worth noting.

(1) z, d, and F_{ts} are defined as positive in the direction normal and away from the surface (see figure 6.1). If the force is repulsive i.e., $F_{ts} > 0$, and largest when the tip is close to the surface, the largest values of $F_{ts}z$ are to be obtained for negative values of cantilever deflection z. In such cases the virial is negative, i.e., $\langle F_{ts}z\rangle < 0$. The implication is that the frequency shift $\frac{\Delta\omega_0}{\omega_0}$ is positive, i.e., $\frac{\Delta\omega_0}{\omega_0} > 0$. Namely, the curve should shift toward larger values of ω.

(2) If the force is attractive instead, i.e., $F_{ts} < 0$, and largest when the tip is close to the surface, the largest values of $F_{ts}z$ are to be obtained for negative values of cantilever deflection z, i.e., $z < 0$. In such cases the virial is positive, i.e., $\langle F_{ts}z\rangle > 0$. The implication is that the frequency shift $\frac{\Delta\omega_0}{\omega_0}$ is negative, i.e., $\frac{\Delta\omega_0}{\omega_0} < 0$. Namely, the curve should shift toward larger values of ω.

(3) From equation (6.48), the effective natural frequency of the oscillator is found to be

$$\omega_0 + \Delta\omega_0 = \omega_0 - \omega_0\frac{1}{kA^2}V$$

$$\omega_0' = \omega_0 - \omega_0\frac{1}{kA^2}V$$

$$\omega_0' = \omega_0 \left[1 - \frac{1}{kA^2}V\right] \tag{6.49}$$

Since $\varphi = 90$ when $\omega = \omega_0'$, the phase shift φ should follow the same distortions as the amplitude.

(4) When there are attractive and repulsive forces contained in F_{ts}, the distortions should be double. Namely, and as discussed when interpreting the figures 5.2–5.7, attractive forces pull the curve to lower values of ω and repulsive forces pull it to higher values.

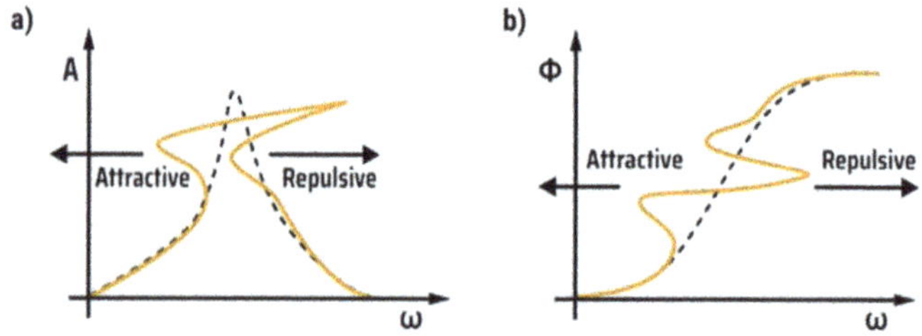

Figure 6.8. Nonlinear case. Illustration of a standard frequency sweep showing the (a) amplitude A and (b) phase ϕ response where the origin of the distortions is shown in terms of whether nonlinear forces are attractive or repulsive. Attractive forces pull the curves to lower frequencies while repulsive forces pull them towards higher frequencies (reproduced from figure 5.4).

The reader can follow the interpretation of equation (6.48) given in the above four points by inspecting figure 5.4 (reproduced below as figure 6.8).

Giessibl already provided the first hint that only conservative forces affect the effective natural frequency ω_0' when he showed that equation (6.48) could be derived from the Hamilton–Jacobi formalism. Such formalism accounts for conservative forces. San Paulo and García derived the virial expression in 2001 and showed that the amplitude and phase curves distort in a way that agrees with equation (6.48) but they felt short of writing down the relation explicitly. This relationship can be found by combining equations (6.47) and (6.48). To our knowledge the first to exploit this relationship were Katan *et al* in 2008 [19, 20]. The authors showed that a general expression for force reconstruction employed in FM AFM, since Sader and Jarvis first presented them in 2004 [21, 22], could be employed in AM AFM by obtaining the frequency shift from the cosine of the angle.

The combination of equations (6.47) and (6.48) gives ($\omega = \omega_0$)

$$\frac{\Delta\omega_0}{\omega_0} = -\frac{1}{kA^2}V$$

$$V = -\frac{1}{2}F_0 A\cos\phi$$

$$\frac{\Delta\omega_0}{\omega_0} = -\frac{1}{kA^2}\left[-\frac{1}{2}F_0 A\cos\phi\right]$$

$$F_0 = \frac{kA_0}{Q} = kA_D$$

$$\frac{\Delta\omega_0}{\omega_0} = \frac{1}{2}\frac{A_0}{AQ}\cos\phi$$

$$\frac{\Delta\omega_0}{\omega_0} = \frac{1}{2}\frac{A_D}{A}\cos\phi \tag{6.50}$$

An expression for any ω will be given in the next chapter.

6.2.3 Amplitude modulation (AM) AFM

In AM AFM A_D, A, and φ are experimentally found. More thoroughly A_D is set by the user, for a given Q, by setting a drive force that produces the required A_0. Then A must be set as a target amplitude to track the surface. The phase shift φ results from the interaction and is a 'free' parameter that provides information about the tip–sample force F_{ts}. In particular, in AM AFM the frequency shift $\frac{\Delta\omega_0}{\omega_0}$ is responsible for the information provided by the phase shift. This is because in order to reach a given amplitude A in AM AFM while driving at a constant ω (this is typical operation in AM AFM), $\frac{\Delta\omega_0}{\omega_0}$ must vary, i.e., the $A(\omega)$ curve must distort.

6.2.4 Frequency modulation (FM) AFM

In FM AFM A_D, and φ are set as a target by the user. In particular φ is set by tracking ω_0'. Here, a frequency shift is set as a target, if the amplitude A is also set to be constant, the system is then tracking a constant virial. This can be shown by referring to equation (6.48). In FM AFM a constant A can be targeted by tracking ω_0' and simultaneously varying the drive, i.e., F_0, to reach a target A. The distortion of the curve is the same in AM and FM since the phenomenon explored is exactly the same, i.e., a cantilever with a sharp tip vibrating near a surface. It is only the parameters being tracked that vary from one method to another.

For more information on AM and FM AFM we refer the reader to the literature [20, 23, 24].

References

[1] Garcia R and San Paulo A 1999 Attractive and repulsive tip–sample interaction regimes in tapping-mode atomic force microscopy *Phys. Rev.* B **60** 4961

[2] Hamaker H C 1937 The London–van der Waals attraction between spherical particles *Physica* **4** 1058–72

[3] Gadelrab K R, Santos S and Chiesa M 2013 Heterogeneous dissipation and size dependencies of dissipative processes in nanoscale interactions *Langmuir* **29** 2200–6

[4] Israelachvili J 1991 *Intermolecular and Surface Forces* 2nd edn (New York: Academic)

[5] Paulo Á S and García R 2002 Unifying theory of tapping-mode atomic-force microscopy *Phys. Rev.* B **66** 041406

[6] Tamayo J and Garcia R 1996 Deformation, contact time, and phase contrast in tapping mode scanning force microscopy *Langmuir* **12** 4430–5

[7] Paulo Á S and García R 2001 Tip-surface forces, amplitude, and energy dissipation in amplitude-modulation (tapping mode) force microscopy *Phys. Rev.* B **64** 193411

[8] Garcia R and San Paulo A 2000 Dynamics of a vibrating tip near or in intermittent contact with a surface *Phys. Rev.* B **61** 13381–4

[9] Israelachvili J N 2005 *Intermolecular and Surface Forces* (London: Elsevier Academic)

[10] Tamayo J and Garcia R 1998 Relationship between phase shift and energy dissipation in tapping-mode scanning force microscopy *Appl. Phys. Lett.* **73** 2926–8

[11] Tamayo J 1999 Energy dissipation in tapping-mode scanning force microscopy with low quality factors *Appl. Phys. Lett.* **75** 3569–71

[12] Cleveland J P, Anczykowski B, Schmid A E and Elings V B 1998 Energy dissipation in tapping-mode atomic force microscopy *Appl. Phys. Lett.* **72** 2613–5

[13] Anczykowski B, Gotsmann B, Fuchs H, Cleveland J P and Elings V B 1999 How to measure energy dissipation in dynamic mode atomic force microscopy *Appl. Surf. Sci.* **140** 376–82

[14] Gadelrab K R, Santos S, Souier T and Chiesa M 2012 Disentangling viscosity and hysteretic dissipative components in dynamic nanoscale interactions *J. Phys. D: Appl. Phys.* **45** 012002

[15] Clausius R Xvi 1870 On a mechanical theorem applicable to heat *Lond. Edinb. Dublin Philos. Mag. J. Sci.* **40** 122–7

[16] Goldstein H, Poole C and Safko J L 2001 *Classical Mechanics* (London: Pearson)

[17] Wikipedia 2022 Virial Theorem wikipedia.org https://en.wikipedia.org/wiki/Virial_theorem.

[18] Giessibl F J 1997 Forces and frequency shifts in atomic-resolution dynamic-force microscopy *Phys. Rev.* B **56** 16010

[19] Katan A J and Oosterkamp T H 2008 Measuring hydrophobic interactions with three-dimensional nanometer resolution *J. Phys. Chem.* C **112** 9769–76

[20] Katan A J, Es M H and Oosterkamp T H 2009 Quantitative force versus distance measurements in amplitude modulation afm: a novel force inversion technique *Nanotechnology* **20** 165703

[21] Sader J E and Jarvis S P 2004 Accurate formulas for interaction force and energy in frequency modulation force spectroscopy *Appl. Phys. Lett.* **84** 1801–3

[22] Sader J E, Uchihashi T, Higgins M J, Farrell A, Nakayama Y and Jarvis S P 2005 Quantitative force measurements using frequency modulation atomic force microscopy—theoretical foundations *Nanotechnology* **16** 94

[23] García R and Perez R 2002 Dynamic atomic force microscopy methods *Surf. Sci. Rep.* **47** 197–301

[24] Giessibl F J 2003 Advances in atomic force microscopy *Rev. Mod. Phys.* **75** 949

IOP Publishing

Oscillations
Theory and applications in AFM

Tuza Adeyemi Olukan, Sergio Santos, Lamiaa Sami Elsherbiny and Matteo Chiesa

Chapter 7

Expanding the expressions of energy dissipation and virial

7.1 Virial, energy dissipation, and harmonics

In this chapter several concepts already developed will be discussed in detail to clarify some technical aspects of the formalism so far.

The derivation of the virial V and the energy dissipation terms E_{dis} has been discussed so far from a point of view of physical phenomena. The virial V has been computed by invoking the virial theorem typically exploited in statistical and classical mechanics [1]. In short, the virial theorem claims that the virial of the force is

$$V = \langle F_{ts}z \rangle = \frac{1}{T}\int_0^T F_{ts}z\,dt \tag{7.1}$$

where only the first harmonic of z is considered when computing V. This can be shown by writing z as a Fourier series expansion in its amplitude-phase form. Namely,

$$z = z_0 + A_1 \cos\left(\omega t - \phi_1\right) + A_2 \cos\left(\omega t - \phi_2\right) + \cdots \tag{7.2}$$

So far we have identified A_1 with A and ϕ_1 with ϕ. Then, the virial has been computed so far as follows:

$$V = \langle F_{ts}z \rangle \equiv \frac{1}{T}\int_0^T F_{ts}A_1 \cos\left(\omega t - \phi_1\right)dt \tag{7.3}$$

Strictly speaking this (equation (7.3)) would be the virial of the fundamental harmonic A_1. This point is particularly relevant in multifrequency AFM, a technique whereby multiple drive forces, at different frequencies, are employed to excite the microcantilever [2–4]. In particular, the equation of motion of higher modes must be employed [2] when driving at higher frequencies implying that equations (2.1)–(2.3) in chapter 2 must be considered [5].

7.1.1 Dissipative forces

The energy dissipation expression has been defined as the negative of the work done by the force F_{ts} per cycle

$$E_{dis} = -\int_0^T F_{ts}\dot{z}dt \tag{7.4}$$

Again, the derivative of equation (7.2) shows that, strictly speaking, E_{dis} (equation (7.4)) has been computed so far in terms of the first harmonic only. Then

$$E_{dis} = \int_0^T F_{ts}\omega A_1 \sin\left(\omega t - \phi_1\right)dt \tag{7.5}$$

Expressions equations (7.3) and (7.5) show that these terms can be more or less identified with the first Fourier coefficients. In 1999, Dürig [6] used the least-action principle to derive an expression equivalent to that of Giessibl (1997), equation (6.48) by also developing a Fourier series expansion. Giessibl's expression was shown to be derivable by considering the lowest-order harmonic approximation. In 2004, Sader *et al* further proposed to expand F_{ts} in terms of its odd and even components to show that equation (6.48) for the frequency shift (FM AFM), or equivalently, the expression for the virial V by San Paulo and García in AM AFM (equation (6.47)), contains only the odd component of F_{ts} to its first harmonic term. The 'damping' was shown to contain an integral expression with the odd terms only. In 2008 Hu and Raman also made the connection between the viral and energy dissipation and Fourier series by proceeding to expand F_{ts} in terms of the Fourier coefficients. In his PhD thesis, Álvarez Amo computed (equations (2.10) and (2.11) in the thesis [7]) the Fourier coefficients of F_{ts} directly from the equation of motion to then show how these, but for a factor, correspond to the typical Fourier coefficients of a_n (virial or conservative interactions) and b_n (energy dissipation or dissipative interactions).

It is important to note that both the cosine and the sine terms exploited to derive the virial V (equation (6.34)) and E_{dis} (equation (6.15)) contain ωt but also the angle ϕ. This is because the physical interpretation to derive these expressions is based on the work done which contains $\dot{z}$ and the virial which contains z. Hu and Raman expanded on this problem by also claiming that the nonlinear force F_{ts} can be written in terms of even and odd terms as follows [8]:

$$F_{ts}\left(d, \dot{d}\right)\Big]_{odd} = \frac{F_{ts}\left(d, \dot{d}\right) - F_{ts}\left(d, -\dot{d}\right)}{2} \tag{7.6}$$

The expression shows that the odd term of the force is identified with forces that are not identical independently of the sense of motion, i.e., the force depends on whether the tip moves to the surface or away from the surface. Clearly, if the force depends on the sense of motion, energy will be dissipated during each cycle. Equation (7.7) was defined by Hu and Raman in a way that odd (dissipative) forces had to meet the condition in equation (7.7) [8],

$$F_{ts}\left(d,\, \dot{d}\right) = -F_{ts}\left(d,\, -\dot{d}\right) \text{ for odd/dissipative forces} \tag{7.7}$$

Equation (7.7) would imply that the odd term contains the part of the force which depends on the sense of motion but is otherwise symmetrical. This is not strictly true. For example, Santos *et al* [9] showed that the formalism in equation (7.6) is general enough to successfully decouple forces that are not identical independently of the sense of motion (equation (7.8)). The following relation thus applies

$$F_{ts}\left(d,\, \dot{d}\right) \neq -F_{ts}\left(d,\, -\dot{d}\right) \text{ for odd/dissipative forces} \tag{7.8}$$

The results predicted by the expression in equation (7.8) was confirmed by Santos *et al* [9] by numerically solving the equation of motion and showing that the formalism that divides the force between odd and even components, can effectively be used to decouple hysteretic and viscous effects. In short while both hysteretic and viscous forces dissipate, hysteretic forces do not necessarily depend on $\dot{d}$ but on the sense of motion and history alone. Viscous forces on the other hand depend on the absolute value of the velocity and might or might not be symmetrical as equation (7.6) assumes. The description of hysteresis and viscosity given here is illustrated in figure 7.1. The figure shows that there might by hysteretic forces in the region of non-contact, i.e., $d > a_0$, and the region of mechanical contact, i.e., $d \leqslant a_0$. The same is true for viscous forces. Viscous forces are typically modelled in AFM by exploiting a Voigt model [10–12]. Then

$$F_{ts}\left(d,\, \dot{d}\right) = -\eta(d)(R\delta)^{1/2}\dot{d} \quad \delta \geqslant 0 \tag{7.9}$$

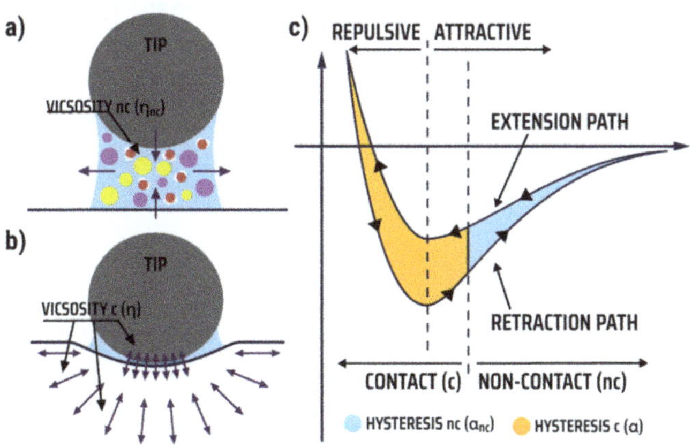

Figure 7.1. Illustrations depicting the mechanisms of viscosity in (a) the nc region and (b) the c region (equation (7.9)). See the literature for details on modelling viscosity in the different regions [12]. (c) Hysteretic processes (equation (7.10)) can be characterized by an increment in force depending on the sense of motion. Reproduced from [12].

where the viscous term might or might not depend on d, η is the viscosity of the tip-sample interaction, R is the tip radius, δ is the deformation and $\dot{d}$ is the velocity. There is a minus sign to indicate that the force opposes the motion. A hysteretic force indicates that there is force depending on the sense of motion and possibly history. For example, capillary interactions are history dependent [13]. A simply hysteretic force can be modelled in AFM as follows:

$$F_{ts}\left(d, \dot{d}\right) = -\alpha F_{AD} \quad \delta \geqslant 0 \text{ and } \dot{d} > 0 \tag{7.10}$$

where α is a coefficient proportional to the force of adhesion FAD. The condition for the even part of the force F_{ts}(even) defined by Hu and Raman (equation (7.11)) was also put to the test by Santos *et al* [9].

$$F_{ts}\left(d, \dot{d}\right)\Big]_{\text{even}} = \frac{F_{ts}\left(d, \dot{d}\right) + F_{ts}\left(d, -\dot{d}\right)}{2} \tag{7.11}$$

It was shown that in the presence of hysteretic components, the condition in equation (7.11) cannot be identified with the conservative force. Rather, the condition gives the average of the force where hysteresis is present. The implication is that equation (7.11), i.e., the even components of the force, cannot be identified at once with conservative forces when hysteretic forces are present. For the same reason equation (7.6) cannot be identified at once with dissipative forces. Rather, if hysteretic forces are present, the force will be contained in both the even and odd components. This is also the case for the equations for the energy dissipation E_{dis}, the virial V and the frequency shift; equations (6.15), (6.34) and (6.38), respectively. Some evidence of this was given by Santos *et al* in 2014 [14]. The dissipative and conservative components can be recovered, however components if different velocities are probed to recover the force and the dissipative and conservative parts are compared for the two different tip velocities. Such a formalism was presented exploiting numerical integration by Santos *et al* in 2012, but it is experimentally challenging and cumbersome [9].

In summary, strictly speaking it is best to use expressions (7.6) and (7.11) to refer to even and odd forces and only roughly identify these terms with conservative and dissipative forces. This point was already emphasized by Sader *et al* [15] in 2005 but it is not a well-known fact. We reproduce the discussion of this point given in 2005 by Sader *et al* in full:

> *However, we emphasize that such connections to 'conservative' and 'dissipative' forces, although common and appealing, can be misleading and ambiguous since the origin of the forces cannot be determined solely by analysing the harmonic motion of the tip in isolation. Indeed, it is entirely possible that the 'conservative' force results from a combination of conservative, dissipative and energy gaining processes over different parts of the oscillation cycle, even though the net energy lost or gained is zero over a complete cycle. Importantly, a formal connection to conservative*

and dissipative forces can only be made if the precise nature and origin of the forces involved is known. Therefore, contrary to convention, it is preferable to refer to these forces simply as 'even' and 'odd' in general, which shall henceforth be adopted [15].

Sader *et al* correctly pointed out that the even component simply requires that 'the net energy lost or gained is zero over a complete cycle.' For example, as corroborated by Santos *et al* [9], in the presence of hysteretic forces (equation (7.9)) the even components produce a 'mean' path, i.e., a zero net energy loss/gain over a cycle, between the approach and retract paths. In terms of the illustration of figure 7.1, this would mean that the average between the approach and retract paths would contribute to the frequency shift/virial in equations.

7.1.2 Sign conventions

The signs in the expressions are defined when the physical phenomenon is modelled. For example, the equation of motion in equation (5.3) can be defined in terms of z, as position (ignoring mean deflection and higher harmonics), as

$$z = A\cos(\omega t - \phi) \tag{7.12}$$

The virial V and energy dissipation E_{dis} expressions equations (6.15) and (7.1) can then be defined

$$V = \langle F_{ts} \cdot z \rangle = \frac{1}{T}\int_0^T F_{ts} A\cos(\omega t - \phi)dt \tag{7.13}$$

$$E_{dis} = -\int_0^T F_{ts}\dot{z}dt \tag{7.14}$$

(1) Since ϕ has physical significance in the driven oscillator, the minus sign in equation (7.12) implies that the angle obtained as the difference in phase from the response and the drive will be taken in absolute terms. Physically, the response z lags in relation to the drive $F_0\cos(\omega t)$ (see figure 6.8).
(2) While $\sin(-\phi) = -\sin\phi$, the minus sign is already considered in equation (7.12). Thus, with the above convention, $\sin\phi \geqslant 0$ always, since $0° \leqslant \phi \leqslant 180°$.
(3) The solution of the equation of motion is equivalent when defining z as

$$z = A\sin(\omega t - \phi) \tag{7.15}$$

Differences again arise in terms of signs since $\dot{z}$ is negative when the cosine is used instead for z. Arvind Raman [16] defines z as q and develops his theory starting from the sine as in equation (7.15). Still his results are equivalent as expected.

With equation (7.12) and using the convention in equation (7.14)

$$z = A\cos(\omega t - \phi)$$

$$E_{dis} = -\int_0^T F_{ts}\dot{z}dt$$

$$E_{dis} = -\int_0^T F_{ts}(-\omega)A\sin(\omega t - \phi)dt$$

$$E_{dis} = \omega A\int_0^T F_{ts}\sin(\omega t - \phi)dt \tag{7.16}$$

Using the convention of Álvarez Amo [7]

$$z = A\cos(\omega t + \phi) \tag{7.17}$$

$$E_{dis} = \int_0^T F_{ts}\dot{z}dt \tag{7.18}$$

$$E_{dis} = \int_0^T F_{ts}(-\omega)A\sin(\omega t + \phi)dt$$

$$E_{dis} = -\omega A\int_0^T F_{ts}\sin(\omega t + \phi)dt \tag{7.19}$$

where it is to be noted that ϕ is negative, i.e., $\phi \leqslant 0$, because of the definition of z in equation (7.17). The derivation is the same as that leading to equation (6.26).

$$E_{dis} = \int_0^T F_{ts}\dot{z}dt$$

$$E_{dis} = \int_0^T \left[-F_0\cos\omega t + m\frac{d^2\dot{z}}{dt^2} + \frac{m\omega_0}{Q}\frac{dz}{dt} + kz\right]\dot{z}dt$$

$$E_{dis} = I_1 + I_2 + I_3 + I_4 \tag{7.20}$$

Note that if the integrals of the four terms on the right cancel out $E_{dis} = 0$ as before. Each of the integrals can be easily solved by considering orthogonality

$$I_2 = I_4 = 0 \tag{7.21}$$

$$I_1 = -\int_0^T F_0\cos\omega t\dot{z}dt$$

$$I_1 = \omega F_0 A\int_0^T \sin(\omega t + \phi)\cos\omega t dt$$

$$I_1 = \omega F_0 A\int_0^T [\sin(\omega t)\cos\phi + \cos(\omega t)\sin\phi]\cos\omega t dt$$

$$I_1 = \omega F_0 A\int_0^T \cos^2(\omega t)\sin\phi dt$$

$$I_1 = \omega F_0 A \sin\phi \frac{T}{2}$$

$$I_1 = \pi F_0 A \sin\phi \tag{7.22}$$

$$I_3 = \int_0^T \frac{m\omega_0}{Q} \dot{z}^2 dt$$

$$I_3 = \frac{m\omega_0}{Q} A^2 \omega^2 \int_0^T \sin^2(\omega t + \phi) dt$$

$$I_3 = \frac{m\omega_0}{Q} A^2 \omega^2 \left[\frac{T}{2} [\sin^2\phi + \cos^2\phi] \right]$$

$$I_3 = \frac{m\omega_0}{Q} \omega\pi A^2 \tag{7.23}$$

Finally, combining equations (7.20)–(7.23), equation (7.24) follows

$$E_{dis} = \pi F_0 A \sin\phi + \frac{m\omega_0}{Q} \omega\pi A^2, \; z = A\cos(\omega t + \phi), \quad E_{dis} = \int_0^T F_{ts} \dot{z} dt \tag{7.24}$$

$$E_{dis} = \pi F_0 A \sin\phi + \frac{m\omega_0}{Q} \omega\pi A^2, \; z = A\cos(\omega t + \phi), \quad E_{dis} = -\int_0^T F_{ts} \dot{z} dt \tag{7.25}$$

Equations (7.24) and (7.25) are equivalent and only the definitions change. The sign convention however needs to be taken into account. Comparing (7.24) with equation (7.25), obtained from the convention in (7.16), shows several issues with signs.

First, the second term on the right of (7.24) adds to the first term on the right.

Second $\sin\phi \leqslant 0$ in equation (7.24) since $\phi \leqslant 0$. This is because the response lags the drive.

Third, the first term on the right in equation (7.24) must be, from considerations of energy conservation, larger in absolute terms than the second term on the right. The result is that

$$E_{dis} = \pi F_0 A \sin\phi + \frac{m\omega_0}{Q} \omega\pi A^2 < 0 \tag{7.26}$$

The interpretation of (7.26) is that the work done by dissipative forces is negative. In the definition of equation (6.15) leading to equation (6.26) (also (7.25)), the energy dissipated is positive since the definition was for the negative of the work done, i.e., energy dissipated, rather than for the work done. While this discussion on signs leads to equivalent results, it is important to maintain the signs consistent when developing the equations in the next chapters.

At $\omega = \omega_0$, equation (7.26) gives

$$E_{dis}(d) = \frac{\pi k A_0 A(d)}{Q} \left[\sin(\phi(d)) + \frac{A(d)}{A_0} \right] \tag{7.27}$$

where it is explicitly stated that E_{dis}, ϕ, and A depend on d. Strictly speaking equation (7.27) should be defined as work done per cycle. Again equations (6.32) and (7.27) are equivalent but the signs are inverted because of the definition of E_{dis}.

Finally, note that Arvind Raman, in his course, obtains an expression like that in equation (7.25) for E_{dis} and an expression like that in equation (6.46) for V when defining

$$z = A \sin(\omega t - \phi) \tag{7.28}$$

This is because Raman defines E_{dis} as the negative of the work done in agreement with the definition in equation (6.15) leading to (7.25). See nanoHUB-U Fundamentals of AFM L2.4 by Arvind Raman [16] (see equations (1) and (2) there).

7.2 The virial and energy dissipation and the frequency shift

7.2.1 The tangent of the phase shift φ in the nonlinear theory

The nonlinear theory will be further developed in this chapter in order to derive expressions for the frequency shift of the 'natural' resonant frequency, to find the transfer function of the nonlinear system to express the perturbed amplitude A in terms of E_{dis} and V.

Using equation (6.46) for V and equation (7.24) for E_{dis} (here (7.29))

$$z = A \cos(\omega t + \phi)$$

$$V = -\frac{1}{2}F_0 A \cos\phi + \frac{1}{2}A^2[k - \omega^2 m] \tag{7.29}$$

$$E_{dis} = \frac{m\omega_0}{Q}\omega\pi A^2 + F_0 A\pi \sin\phi \tag{7.30}$$

Since the virial V is a function of the cosine of the phase shift ($-180° \leqslant \phi \leqslant 0°$), the virial can be both positive and negative. The sine is always negative for this range ($-180° \leqslant \phi \leqslant 0°$). The energy dissipation expression must always be positive by definition. Nevertheless we have used the convention of Álvarez Amo (see the chapter on sign convention) to express E_{dis} in terms of the work done by the F_{ts} force (equation (7.29)). Next an expression for $\tan\phi$ based on the thesis of Álvarez Amo [7] is developed.

Rearranging equation (7.29)

$$2V = A^2 m\left(\omega_0^2 - \omega^2\right) - F_0 A \cos\phi$$

$$2V = A^2 k\left(1 - \left[\frac{\omega}{\omega_0}\right]^2\right) - F_0 A \cos\phi \tag{7.31}$$

An approximation for F_0 is now employed to simplify equation (7.31)

$$F_0 = \frac{kA_0}{Q} \tag{7.32}$$

Equation (7.32) assumes that $\omega = \omega_0$. See, for example, the linear expression for A in equation (5.7). Dividing equation (7.31) by (7.32)

$$V\frac{2Q}{kA_0} = \frac{Q}{A_0}\left[1 - \left(\frac{\omega}{\omega_0}\right)^2\right]A^2 - A\cos\phi \tag{7.33}$$

Rearranging the expression for E_{dis} in equation (7.29) (again following Álvarez Amo's development)

$$E_{dis} = \frac{m\omega_0}{Q}\omega\pi A^2 + F_0 A\pi \sin\phi$$

$$E_{dis} = \frac{k}{Q}\frac{\omega}{\omega_0}\pi A^2 + F_0 A\pi \sin\phi \tag{7.34}$$

Finally, dividing by F_0 with the same assumption of equation (7.32)

$$E_{dis}\frac{Q}{\pi k A_0} = \frac{1}{A_0}\left(\frac{\omega}{\omega_0}\right)A^2 + A\sin\phi \tag{7.35}$$

Equations (7.33) and (7.35) can be rearranged to give

$$A\cos\phi = \frac{Q}{A_0}\left[1 - \left(\frac{\omega}{\omega_0}\right)^2\right]A^2 - \frac{2Q}{kA_0}V$$

$$A\sin\phi = E_{\text{dis}}\frac{Q}{\pi k A_0} - \frac{\mu}{A_0}A^2 \text{ where } \mu = \frac{\omega}{\omega_0} \tag{7.36}$$

Maintaining the cosine and the sine on the left-hand side of the expressions is useful to compute $\tan\varphi$ but the expressions can also be rearranged with the terms V and E_{dis} on the left-hand side. Then

$$V\frac{2Q}{kA_0} = \frac{Q}{A_0}\left[1 - \left(\frac{\omega}{\omega_0}\right)^2\right]A^2 - A\cos\phi \tag{7.37}$$

$$E_{\text{dis}}\frac{Q}{\pi k A_0} = \frac{\mu}{A_0}A^2 + A\sin\phi \tag{7.38}$$

The following definitions are then proposed by Álvarez Amo [7]

$$\varepsilon = -E_{dis}\frac{Q}{\pi K A_0} \tag{7.39}$$

$$\nu = V\frac{2Q}{kA_0} \tag{7.40}$$

The negative sign in the expression in equation (7.39) by Álvarez Amo effectively redefines E_{dis} as the energy dissipation. That is, the integral is redefined as equation (7.16), reproduced here to support this discussion

$$E_{dis} = -\int_0^T F_{ts}\dot{z}dt$$

$$\varepsilon = E_{dis}\frac{Q}{\pi KA_0} = -\frac{Q}{\pi KA_0}\int_0^T F_{ts}\dot{z}dt \quad (7.41)$$

Combining equations (7.38) and (7.39)

$$E_{dis} = -\frac{\pi kA_0A}{Q}\left[\sin\,\phi + \mu\frac{A}{A_0}\right] \quad (7.42)$$

The above expression still assumes that $z = A\cos(\omega t + \varphi)$ but now we are speaking of energy dissipated rather than work done by the force F_{ts}. The development of the formalism continues by defining, from the combination of equations (7.36) and (7.39)

$$A\sin\,\phi = -\varepsilon - \mu\frac{A^2}{A_0} \quad (7.43)$$

Combining equations (7.36) and (7.40)

$$\varepsilon = -\left(\frac{\mu}{A_0}A^2 + A\sin\,\phi\right) \quad \phi \leqslant 0 \quad (7.44)$$

In summary, since the assumptions leading to equation (7.44) are that $\phi \leqslant 0$ and that $F_0 = \frac{kA_0}{Q}$ (this is an approximation that assumes $\omega = \omega_0$), and since in absolute terms $A\sin\,\phi > \frac{\mu}{A_0}A^2$, using ε rather than E_{dis} effectively leads to talking about E_{dis} as a positive term. Then

$$E_{dis} = -\frac{\pi kA_0A}{Q}\left[\sin\,\phi + \mu\frac{A}{A_0}\right] \quad (7.45)$$

The above expression still assumes that $z = A\cos(\omega t + \varphi)$ but now we are speaking of energy dissipated rather than work done by the force F_{ts}. The development of the formalism continues by defining, from the combination of equations (7.36) and (7.39)

$$A\sin\,\phi = -\varepsilon - \mu\frac{A^2}{A_0} \quad (7.46)$$

Combining equations (7.36) and (7.40)

$$A\cos\,\phi = \frac{Q}{A_0}[1 - \mu^2]A^2 - \nu \quad (7.47)$$

The discussion continues by assuming that $\omega = \omega_0$. This assumption implies that $F_0 = \frac{kA0}{Q}$ and thus the assumption in equations (7.33) and (7.35) hold. Then ($\omega = \omega_0$)

$$A \cos \phi = -\nu \tag{7.48}$$

$$A \sin \phi = -\varepsilon - \frac{A^2}{A_0} \tag{7.49}$$

Equations (7.48) and (7.49) can now be used to compute $\tan \phi$ following the convention of Álvarez Amo [7]

$$\tan \phi = \frac{A^2}{A_0 \nu} \tag{7.50}$$

Writing equation (7.50) in terms of V ($\omega = \omega_0$)

$$\tan \phi = \frac{\frac{A^2}{A_0}}{V \frac{2Q}{kA_0}}$$

$$\tan \phi = \frac{kA^2}{2Q} \frac{1}{V} \tag{7.51}$$

The virial V can be written in terms of the cosine of the phase shift. Then ($\omega = \omega_0$)

$$\tan \phi = \frac{kA^2}{2Q} \cdot \frac{1}{-\frac{1}{2} F_0 A \cos \phi}$$

$$\tan \phi = \frac{kA^2}{2Q} \cdot \frac{-1}{\frac{kA_0}{2Q} A \cos \phi}$$

$$\tan \phi = \frac{-A}{A_0 \cos \phi} \tag{7.52}$$

It is trivial to show that equation (7.52) holds for equations (7.17) and (7.18). Álvarez Amo's convention where $-180° \leqslant \varphi \leqslant 0°$

$$z = A \cos (\omega t + \phi) \tag{7.53}$$

$$E_{dis} = \int_0^T F_{ts} \dot{z} dt \tag{7.54}$$

Equation (7.52) simply expresses that $E_{dis} = 0$. This can be shown by writing from equation (7.52)

$$\tan \phi \cos \phi = -\frac{A}{A_0}$$

$$\sin\phi = -\frac{A}{A_0} \tag{7.55}$$

Equations (7.52) and (7.55) result from the consideration of the same phenomenon, namely, tracking a shift in natural frequency ω_0 due to conservative, i.e., or strictly speaking, even, forces. This can be shown by considering the energy dissipation equation at $\omega = \omega_0$, from Álvarez Amo's convention

$$E_{dis} = -\frac{\pi k A_0 A}{Q}\left[\sin\phi + \mu\frac{A}{A_0}\right]$$

Combining equations (7.45) and (7.55) at $\omega = \omega_0(\mu = 1)$

$$E_{dis} = -\frac{\pi k A_0 A}{Q}\left[-\frac{A}{A_0} + \frac{A}{A_0}\right] = 0 \tag{7.56}$$

7.2.2 The solution for tan ϕ for any ω

The expression $\tan\phi$ for any ω can be found directly from equations (7.29) and (7.30) without invoking the assumption that $F0 = \frac{kA0}{Q}$ that holds only for $\omega = \omega_0$. Rearranging the equations

$$A\cos\phi = \frac{A^2}{F_0}[k - \omega^2 m] - \frac{2}{F_0}V \tag{7.57}$$

$$A\sin\phi = \frac{1}{\pi F_0}E_{dis} - \frac{k}{QF_0}\mu A^2 \quad \mu = \frac{\omega}{\omega_0} \tag{7.58}$$

Then,

$$\tan\phi = \frac{\frac{1}{\pi F0}E_{dis} - \frac{k}{QF0}\mu A^2}{\frac{A^2}{F0}[k - \omega^2 m] - \frac{2}{F0}V} \tag{7.59}$$

The result in equation (7.51) by Álvarez Amo can be found by setting $\omega = \omega_0$, for which $F_0 = \frac{kA0}{Q}$, and the condition that $E_{dis} = 0$ in equation (7.59). Then

$$\tan\phi = \frac{kA^2}{2Q}\frac{1}{V} \tag{7.60}$$

The above expression is the same as that in equation (7.51). Nevertheless this is because $\omega = \omega_0$. Strictly speaking, the correct results for any ω are those in equations (7.57)–(7.60) since Álvarez Amos's approximation for F_0, i.e., $F_0 = \frac{kA0}{Q}$, is valid only at $\omega = \omega_0$.

7.2.3 Comparison between sign conventions for tan φ

The convention established by equations (6.26), (6.46) and (7.14)–(7.16) can be shown to lead to the same results as those by Álvarez Amo's convention (equation (7.60)). The derivation is carried out here in detail. From the alternative definition

$$Az = A\cos(\omega t - \phi)$$

$$E_{dis} = -\int_0^T F_{ts}\dot{z}dt$$

$$E_{dis} = \pi F_0 A \sin\phi - \frac{m\omega_0}{Q}\omega\pi A^2$$

$$V = \frac{1}{2}A^2[k - \omega^2 m] - \frac{1}{2}F_0 A\cos\phi$$

The sine and cosine functions can now be expressed as

$$A\cos\phi = \frac{A^2}{F_0}[k - \omega^2 m] - \frac{2}{F_0}V \tag{7.61}$$

$$A\sin\phi = \frac{1}{\pi F_0}E_{dis} + \frac{m\omega_0}{QF_0}\omega A^2 \tag{7.62}$$

The expression for tan ϕ gives

$$\tan\phi = \frac{\frac{1}{\pi F_0}E_{dis} + \frac{k}{QF_0}\mu A^2}{\frac{A^2}{F_0}[k - \omega^2 m] - \frac{2}{F_0}V} \quad \text{where } z = A\cos(\omega t - \phi),\ E_{dis} = -\int_0^T F_{ts}\dot{z}dt \tag{7.63}$$

The expression in equation (7.63) is equivalent to that in (7.59) (Álvarez Amo) since cos ϕ is an even function and symmetrical around $\phi = 0$. Compare (7.63) with (7.59) below

$$\tan\phi = \frac{\frac{1}{\pi F_0}E_{dis} - \frac{k}{QF_0}\mu A^2}{\frac{A^2}{F_0}[k - \omega^2 m] - \frac{2}{F_0}V} \quad \text{where } z = A\cos(\omega t + \phi),\ E_{dis} = \int_0^T F_{ts}\dot{z}dt$$

In summary,

(1) In equation (7.59) the term containing $E_{dis} < 0$ because of the definition given to compute E_{dis}.
(2) In equation (7.59) the two terms on the numeration are negative and the numerator is positive in both equations (7.59) and (7.63).
(3) Because of the above two points the phase shift in equation (7.59) is negative and since the tan function is odd, $-\tan(\phi) = \tan(-\phi)$.
(4) By definition, the term in the numerator in (7.63) is positive and the term in the denominator is the same as in (7.59) because cos ϕ is an even function, $\cos(-\phi) = \cos\phi$.

Finally, either equation (7.59) or equation (7.63) can be used to compute the tangent of ϕ or E_{dis} provided

(1) The term E_{dis} is interpreted as energy dissipation in (7.63) and as work done W by F_{ts} in (7.59). Then

$$E_{dis} = -W \tag{7.64}$$

(2) The sign of the phase shift ϕ in (7.63) is considered as negative, i.e., the response lags the drive, and the phase shift in (7.59) is considered as positive and the sign is defined already when defining z.

For the condition $\omega = \omega_0$ and $E_{dis} = 0$, equation (7.63) gives

$$\tan\phi = -\frac{kA^2}{2Q}\frac{1}{V} \quad \text{where } z = A\cos(\omega t - \phi),\ E_{dis} = -\int_0^T F_{ts}\dot{z}dt \tag{7.65}$$

Again, equations (7.60) and (7.65) only differ in terms of a negative sign because the negative sign is already accounted for when defining z in equation (7.65). Writing $\tan\phi$ in terms of the cosine function

$$\tan\phi = -\frac{kA^2}{2Q}\frac{1}{\left[-\frac{1}{2}AF_0\cos\phi\right]}$$

$$\tan\phi = \frac{A}{A_0\cos\phi} \quad \text{if } \omega = \omega_0,\ F_0 = \frac{kA_0}{Q} \tag{7.66}$$

The difference in sign is encountered again in equation (7.66) as compared with equation (7.52).

7.2.3.1 The (natural) frequency shift in the nonlinear theory

The natural frequency of the linear oscillator ω_0 does not coincide with the resonant frequency ω_r. The resonant frequency, sometimes called effective frequency in nonlinear theory, is the sum of both ω_0 and $\Delta\omega_0$.

$$\omega_r = \omega_{eff} = \omega_0' = \omega_0 + \Delta\omega_0 \tag{7.67}$$

7.2.4 Deriving the 'natural' frequency shift from the assumption of a weakly perturbed oscillator

The derivation first presented by Giessibl in 1997 can be derived directly from the results of the linear theory. A similar derivation is provided by Álvarez Amo in his thesis [7]. Here some details will be given regarding the sign convention and solutions.

Using equation (5.8) from the driven oscillator, one can deduce that

$$\tan\phi = \frac{\frac{\omega\omega_0}{Q}}{\omega_0^2 - \omega^2}$$

$$\tan\phi = \frac{1}{Q\left[\frac{\omega_0}{\omega} - \frac{\omega}{\omega_0}\right]} \tag{7.68}$$

Equation (7.69) is a main source of issues when it comes to signs. In principle, the phase shift is negative, implying that either

$$\tan\phi = \frac{1}{Q\left[\frac{\omega_0}{\omega} - \frac{\omega}{\omega_0}\right]} \quad \text{linear theory } z = A\cos(\omega t - \phi) \tag{7.69}$$

or

$$\tan\phi = -\frac{1}{Q\left[\frac{\omega_0}{\omega} - \frac{\omega}{\omega_0}\right]} \quad \text{linear theory } z = A\cos(\omega t + \phi) \tag{7.70}$$

In equation (7.70) the sign is defined in terms of z whereas the sign is conserved in φ in (7.71). We recall that the term $\tan\varphi$ in the nonlinear theory results in (nonlinear theory, $\omega = \omega_0$)

$$\tan\phi = \frac{kA^2}{2Q}\frac{1}{V} \quad \text{where } z = A\cos(\omega t + \phi),\ E_{dis} = \int_0^T F_{ts}\dot{z}dt$$

$$\tan\phi = -\frac{\mathrm{k}A^2}{2Q}\frac{1}{V} \quad \text{where } z = A\cos(\omega t - \phi),\ E_{dis} = -\int_0^T F_{ts}\dot{z}dt$$

The weakly perturbed resonant frequency shift ω_0' can be derived from equation (7.69) or equation (7.70). Solving equation (7.69) for ω_0,

$$\omega_0 = \frac{\omega}{2Q\tan\phi}\left[\sqrt{1 + 4Q^2\tan^2\phi} + 1\right] \quad \text{linear theory} \tag{7.71}$$

Solving equation (7.70) for ω_0,

$$\omega_0 = \frac{\omega}{2Q\tan\phi}\left[\sqrt{1 + 4Q^2\tan^2\phi} - 1\right] \quad \text{linear theory} \tag{7.72}$$

The solution of (7.69) and (7.70) can be easily found (equations (7.71) and (7.72)) by exploiting software such as Matlab. For example,

Matlab script to solve (7.69)

```
syms theta w w0 Q
F=tan(theta)+(-w*w0/Q)/(w0^2−w^2);
Sol = solve(F,w0)
```

Matlab script to solve (7.70)

```
syms theta w w0 Q
F=tan(theta)+(w*w0/Q)/(w0^2-w^2);
Sol = solve(F,w0)
```

The new natural frequency can be defined by using equations (7.71) and (7.72) but replacing tan ϕ as defined by equation (7.60)

$$\omega_0' = \frac{\omega}{2Q[\tan\phi]_{\text{nonlinear}}}\left[\sqrt{1+4Q^2\ [\tan^2\phi]_{\text{nonlinear}}} - 1\right] \quad \text{nonlinear theory} \tag{7.73}$$

$$\omega_0' = \frac{\omega}{2Q[\tan\phi]_{\text{nonlinear}}}\left[\sqrt{1+4Q^2\ [\tan^2\phi]_{\text{nonlinear}}} + 1\right] \quad \text{nonlinear theory} \tag{7.74}$$

From now on the subscript nonlinear is dropped since it is cumbersome.

Equation (7.73) with (7.65) and (7.74) can be rewritten as

$$\omega_0' = \omega\left[\sqrt{1+\frac{1}{4Q^2\tan^2\phi}} - \frac{1}{2Q\tan\phi}\right] \quad \text{nonlinear theory} \tag{7.75}$$

$$\omega_0' = \omega\left[\sqrt{1+\frac{1}{4Q^2\tan^2\phi}} + \frac{1}{2Q\tan\phi}\right] \quad \text{nonlinear theory} \tag{7.76}$$

The expressions in equations (7.75) and (7.76) can be simplified by writing

$$x = 2Q\tan\phi \tag{7.77}$$

A Taylor expansion that ignores higher terms produces, from equation (7.75)

$$\omega_0' = \omega\left[\sqrt{1+\frac{1}{x^2}} - \frac{1}{x}\right]$$

$$\omega_0' \approx \omega\left[1 + \frac{1}{2x^2} - \frac{1}{x}\right] \tag{7.78}$$

Since in both (equivalent) expressions for $\tan\phi$ in the nonlinear theory, i.e., equations (7.60) and (7.65), the virial V is in the denominator and kA^2 is in the numerator, it follows that

$$\omega_0' \approx \omega\left[1 - \frac{1}{x}\right] \quad \text{where } kA^2 \gg V \tag{7.79}$$

Combining equation (7.79) with (7.60) and (7.77), implying that $\omega = \omega_0$,

$$\omega_0' \approx \omega_0\left[1 - \frac{1}{2Q\tan\phi}\right] \quad \text{where } \tan\phi = \frac{kA^2}{2Q}\frac{1}{V} \tag{7.80}$$

$$\omega_0' \approx \omega_0\left[1 - \frac{V}{kA^2}\right] \tag{7.81}$$

Finally,

$$\frac{\omega_0'}{\omega_0} \approx 1 - \frac{V}{kA^2}$$

$$\frac{\omega_0 + \Delta\omega_0}{\omega_0} \approx 1 - \frac{V}{kA^2}$$

$$\frac{\Delta\omega_0}{\omega_0} \approx -\frac{V}{kA^2} \tag{7.82}$$

Equation (7.82) is equivalent to that provided by Giessibl in 1997 when exploiting the Hamilton Jacobi formalism.

The reader can confirm that a similar derivation leads to equation (7.82) with the correct sign by combining equations (7.76) and (7.65).

Finally, the virial V at $\omega = \omega_0$ is given in AM AFM by

$$V = -\frac{1}{2}F_0 A \cos\phi$$

Then, combining equations (7.82) and (6.47)

$$\frac{\Delta\omega_0}{\omega_0} \approx \frac{1}{2}\frac{F_0}{kA}\cos\phi \tag{7.83}$$

But $F_0 = \frac{kA0}{Q}$ (at $\omega = \omega_0$). Then

$$\frac{\Delta\omega_0}{\omega_0} \approx \frac{1}{2}\frac{A_0}{QA}\cos\phi$$

$$\frac{\Delta\omega_0}{\omega_0} \approx \frac{1}{2}\frac{A_D}{A}\cos\phi \tag{7.84}$$

The result is equivalent to that given in equation (6.50).

7.2.5 Frequency shift and virial

The first derivation of the frequency shift in 1997 by Giessibl [17] was presented as follows (section 6.2.2 under 'the virial of the interaction')

$$\frac{\Delta f_0}{f_0} = -\frac{1}{kA^2}\langle F_{ts} z\rangle$$

$$z = A\cos(\omega t + \phi)$$

where

$$\langle F_{ts} z\rangle = \frac{1}{T}\int_0^T F_{ts} z\, dt$$

The expressions in equations (6.34), (6.48), and (7.53) were derived to compute the frequency shift in FM AFM in 1997. Equation (6.34) is equivalent to the virial V first invoked [18] to derive an expression for the amplitude reduction in AM AFM.

Here we explore alternative methods to write down equations (6.34) and (6.48). For example, Sader and Jarvis [19] write

$$\frac{\Delta\omega_0}{\omega_0} = -\frac{1}{\pi k A}\int_{-1}^{1} F_{ts}\frac{u}{\sqrt{1-u^2}}du \tag{7.85}$$

The expression in equation (7.85) is found by applying a change of variable in equation (6.48) as follows:

$$\langle F_{ts}z\rangle = \frac{1}{T}\int_0^T F_{ts}z dt$$

$$\langle F_{ts}z\rangle = \frac{1}{T}\int_0^T F_{ts}A\cos(\omega t + \phi)dt$$

$$\langle F_{ts}z\rangle = \frac{A}{T}\frac{1}{\omega}\int_0^{2\pi} F_{ts}\cos\varphi d\varphi$$

$$\langle F_{ts}z\rangle = \frac{A}{2\pi}\int_0^{2\pi} F_{ts}\cos\varphi d\varphi \tag{7.86}$$

where

$$\omega = \frac{2\pi}{T},\ \varphi = \omega t + \phi \quad \text{and} \quad d\varphi = \omega dt \tag{7.87}$$

Another change of varible is typically applied

$$\langle F_{ts}z\rangle = \frac{A}{2\pi}\int_0^{2\pi} F_{ts}\cos\varphi d\varphi$$

$$\langle F_{ts}z\rangle = \frac{A}{2\pi}\oint F_{ts}\frac{u}{\pm\sqrt{1-u^2}}du \tag{7.88}$$

where

$$\cos\varphi = u$$

$$-\sin\varphi d\varphi = du$$

$$\sin\varphi = \sqrt{1-u^2} \tag{7.89}$$

For the sake of easing the computation of equation (7.88)

$$\langle F_{ts}z\rangle = \frac{A}{2\pi}\oint F_{ts}\frac{u}{\pm\sqrt{1-u^2}}du \equiv \frac{A}{\pi}\int_{u=-1}^{u=1} F_{ts}\frac{u}{\sqrt{1-u^2}}du \tag{7.90}$$

where equation (7.90) exploits the fact that the 1/2*e* period should be identical to the other half provided the motion of the tip is sufficiently sinusoidal, i.e., $z \propto \cos(\omega t + \phi)$.

Finally, combining equations (6.48) and (7.90)

$$\frac{\Delta\omega_0}{\omega_0} = -\frac{1}{kA^2}\left[\frac{A}{\pi}\int_{u=-1}^{u=1} F_{ts}\frac{u}{\sqrt{1-u^2}}du\right]$$

$$\frac{\Delta\omega_0}{\omega_0} = -\frac{1}{\pi kA}\int_{u=-1}^{u=1} F_{ts}\frac{u}{\sqrt{1-u^2}}du \tag{7.91}$$

where starting from the standard form of the virial V in equation (7.86) the integral in equation (7.91) is found. Equation (7.91) is the expression used by Sader *et al* in equation (7.85).

References

[1] Goldstein H, Poole C and Safko J L 2001 *Classical Mechanics* (London: Pearson)

[2] Rodriguez T and Garcia R 2004 Compositional mapping of surfaces in atomic force microscopy by excitation of the second normal mode of the microcantilever *Appl. Phys. Lett.* **84** 449–551

[3] Lozano J R and Garcia R 2009 Theory of phase spectroscopy in bimodal atomic force microscopy *Phys. Rev.* B **79** 014110

[4] Lozano J R and Garcia R 2008 Theory of multifrequency atomic force microscopy *Phys. Rev. Lett.* **100** 76102–5

[5] Rodríguez T R and García R 2002 Tip motion in amplitude modulation (tapping-mode) atomic-force microscopy: comparison between continuous and point-mass models *Appl. Phys. Lett.* **80** 1646–8

[6] Dürig U 2000 Extracting interaction forces and complementary observables in dynamic probe microscopy *Appl. Phys. Lett.* **76** 1203–5

[7] Amo C 2019 *Microscopía De Fuerzas Bimodal Y No Resonante Para Medir Propiedades Físicas Y Químicas a Escala Nanométrica* (Madrid: Universidad Autónoma de Madrid)

[8] Hu S and Raman A 2008 Inverting amplitude and phase to reconstruct tip–sample interaction forces in tapping mode atomic force microscopy *Nanotechnology* **19** 375704

[9] Santos S, Gadelrab K R, Barcons V, Font J, Stefancich M and Chiesa M 2012 The additive effect of harmonics on conservative and dissipative interactions *J. Appl. Phys.* **112** 124901–8

[10] Tamayo J and Garcia R 1996 Deformation, contact time, and phase contrast in tapping mode scanning force microscopy *Langmuir* **12** 4430–5

[11] Gadelrab K R, Santos S, Souier T and Chiesa M 2012 Disentangling viscosity and hysteretic dissipative components in dynamic nanoscale interactions *J. Phys. D: Appl. Phys.* **45**

[12] Santos S, Barcons V, Verdaguer A, Font J, Thomson N H and Chiesa M 2011 How localized are energy dissipation processes in nanoscale interactions? *Nanotechnology* **22** 345401

[13] Barcons V, Verdaguer A, Font J, Chiesa M and Santos S 2012 Nanoscale capillary interactions in dynamic atomic force microscopy *J. Phys. Chem.* C **116** 7757–6

[14] Santos S, Amadei C A, Tang T-C, Barcons V and Chiesa M 2019 Deconstructing the Governing Dissipative Phenomena in the Nanoscale arXiv:1401.6587 [cond-mat.mes-hall]

[15] Sader J E, Uchihashi T, Higgins M J, Farrell A, Nakayama Y and Jarvis S P 2005 Quantitative force measurements using frequency modulation atomic force microscopy—theoretical foundations *Nanotechnology* **16** 94

[16] Raman A 2014 Atomic force microscopy https://nanohub.org/resources/520

[17] Giessibl F J 1997 Forces and frequency shifts in atomic-resolution dynamic-force microscopy *Phys. Rev.* B **56** 16010

[18] Paulo Á S and García R 2001 Tip-surface forces, amplitude, and energy dissipation in amplitude-modulation (tapping mode) force microscopy *Phys. Rev.* B **64** 193411

[19] Sader J E and Jarvis S P 2004 Interpretation of frequency modulation atomic force microscopy in terms of fractional calculus *Phys. Rev.* B **70** 012303

IOP Publishing

Oscillations

Theory and applications in AFM

Tuza Adeyemi Olukan, Sergio Santos, Lamiaa Sami Elsherbiny and Matteo Chiesa

Chapter 8

Nonlinear amplitude decay, frequency shift and transfer function

8.1 Frequency shift and virial at arbitrary drive frequencies

This chapter discusses the expression for the frequency shift $\frac{\Delta\omega_0}{\omega_0}$ at any drive frequency ω, i.e., when it is possible that $\omega \neq \omega_0$. The general expression for $\tan\phi$ in terms of E_{dis} and V is

$$\tan\phi = \frac{\frac{1}{\pi F_0}E_{dis} - \frac{k}{QF_0}\mu A^2}{\frac{A^2}{F_0}[k - \omega^2 m] - \frac{2}{F_0}V} \quad \text{where } z = A\cos(\omega t + \phi),\ E_{dis} = \int_0^T F_{ts}\dot{z}dt$$

$$\tan\phi = \frac{\frac{1}{\pi F_0}E_{dis} + \frac{k}{QF_0}\mu A^2}{\frac{A^2}{F_0}[k - \omega^2 m] - \frac{2}{F_0}V} \quad \text{where } z = A\cos(\omega t - \phi),\ E_{dis} = -\int_0^T F_{ts}\dot{z}dt$$

It has already been shown that for $\omega = \omega_0 (\mu = 1)$, $E_{dis} = 0$ and $F_0 = \frac{kA_0}{Q}$, and $V = -\frac{1}{2}AF_0\cos(\phi)$, equation (7.59) and (7.63) reduce to

$$\tan\phi = \frac{-A}{A_0\cos\phi} \quad z = A\cos(\omega t + \phi)$$

$$\tan\phi = \frac{A}{A_0\cos\phi} \quad z = A\cos(\omega t - \phi)$$

The general expression for the virial V is the same independently of the definition of z, then

$$V = -\frac{1}{2}F_0 A\cos\phi + \frac{1}{2}A^2[k - \omega^2 m]$$

doi:10.1088/978-0-7503-5809-5ch8

From the weakly perturbed oscillator formalism developed above (combination of equation (7.77) and equation (7.79)), the general expression for ω_0' at any ω can be written as

$$\omega_0' \approx \omega\left[1 - \frac{1}{2Q\tan\phi}\right] \quad \text{where } kA^2 \gg V \quad z = A\cos(\omega t + \phi) \tag{8.1}$$

$$\omega_0' \approx \omega\left[1 + \frac{1}{2Q\tan\phi}\right] \quad \text{where } kA^2 \gg V \quad z = A\cos(\omega t - \phi) \tag{8.2}$$

Again, the two expressions above are equivalent, but the signs must be considered depending on the definition of z.

The theory is next elaborated for the convention in equation (8.2).

Combining equation (7.63) and (8.2), the general natural (resonant) frequency is found

$$\omega_0' \approx \omega\left[1 + \frac{1}{2}\frac{A^2[k - \omega^2 m] - 2V}{\frac{Q}{\pi}E_{dis} + k\mu A^2}\right] \tag{8.3}$$

The expression in equation (6.48) can be recovered from the general equation (8.3) when $\omega = \omega_0(\mu = 1)$ and $E_{dis} = 0$

$$1 + \frac{\Delta\omega_0}{\omega_0} \approx 1 + \frac{1}{2}\frac{(-2V)}{kA^2}$$

$$\frac{\Delta\omega_0}{\omega_0} \approx -\frac{V}{kA^2} \tag{8.4}$$

8.1.1 The nonlinear transfer function

In this section, we look at the nonlinear resonant frequency and dissipation by exploring the linear theory from another point of view. This alternative approach is explored by Arvind Raman in his course ('nanoHUB-U Fundamentals of AFM L2.4: Analytical Theory—Amplitude, Phase, Frequency Shift, Excitation' [1]).

The analysis starts from E_{dis} and V in terms of the sine and cosine functions

$$A\cos\phi = \frac{A^2}{F_0}[k - \omega^2 m] - \frac{2}{F_0}V$$

$$A\sin\phi = \frac{1}{\pi F_0}E_{dis} + \frac{m\omega_0}{QF_0}\omega A^2$$

Pythagoras theorem provides the mean to relate the sine and cosine function above

$$1 = \sin^2\phi + \cos^2\phi \tag{8.5}$$

Combining equations (7.61), (7.62) and (8.5)

$$\frac{kA}{F_0} = \frac{1}{\sqrt{\left(\frac{\omega_0'^2 - \omega^2}{\omega_0^2}\right)^2 + \left(\frac{\omega}{\omega_0 Q'}\right)^2}} \tag{8.6}$$

where

$$\omega_0' = \omega_0 - \omega_0 \frac{V}{kA^2} \tag{8.7}$$

and

$$\frac{1}{Q'} = \frac{1}{Q} + \frac{\omega_0 E_{dis}}{\omega \pi k A^2} \tag{8.8}$$

Equation (8.7) is equivalent to (6.48). Equation (8.8) is the effective Q factor. The advantage of writing a transfer function such as that in equation (8.6) is that it is analogous to the linear transfer function in equation (4.21) in chapter 4.

$$A = \frac{F_0}{\sqrt{m^2(\omega_0^2 - \omega^2)^2 + b^2\omega^2}}$$

$$\frac{kA}{F_0} = \frac{1}{\sqrt{\left(\frac{\omega_0^2 - \omega^2}{\omega_0^2}\right)^2 + \left[\frac{\omega}{\omega_0 Q}\right]^2}} \tag{8.9}$$

Indeed the whole point of rearranging the expressions in equation (7.61) and (7.62) to get to (8.6) is that the term ω_0' seems to be playing the role of ω_0 in the linear theory and Q' that of Q. For this reason the expressions for ω_0' and Q' can be called 'effective', i.e., they take values that make the system behave 'as if' the linear system has such natural frequency and Q values. Compare, for example, the positions in the equations in equations (8.6) and (8.9) above.

Equation (8.10) is simplified at $\omega = \omega_0$

$$\frac{kA}{\frac{kA_0}{Q}} = \frac{1}{\sqrt{\left(1 - \left[\frac{\omega_0'}{\omega_0}\right]^2\right)^2 + \left(\frac{1}{Q'}\right)^2}} \quad \text{where } \frac{\omega_0'}{\omega_0} = 1 - \frac{V}{kA^2} \text{ and } F_0 = \frac{kA_0}{Q} \tag{8.10}$$

Combining equations (8.7), (8.8) and (8.10)

$$\frac{A}{A_D} = \frac{1}{\sqrt{\left(1 - \left[1 - \frac{V}{kA^2}\right]^2\right)^2 + \left(\frac{1}{Q} + \frac{E_{dis}}{\pi k A^2}\right)^2}}$$

$$\frac{A}{A_D} = \frac{1}{\sqrt{\left(1 - \left[1 + \left[\frac{V}{kA^2}\right]^2 - 2\frac{V}{kA^2}\right]\right)^2 + \left(\frac{1}{Q} + \frac{E_{dis}}{\pi k A^2}\right)^2}}$$

$$\frac{A}{A_D} = \frac{1}{\sqrt{\left(\left[\frac{V}{kA^2}\right]^2 - 2\frac{V}{kA^2}\right)^2 + \left(\frac{1}{Q} + \frac{E_{dis}}{\pi kA^2}\right)^2}} \tag{8.11}$$

Considering the case of a weakly perturbed oscillator $V \ll kA^2$ (see equations (8.1) and (8.2))

$$\frac{A}{A_D} \approx \frac{1}{\sqrt{\left(-2\frac{V}{kA^2}\right)^2 + \left(\frac{1}{Q} + \frac{E_{dis}}{\pi kA^2}\right)^2}} \quad \text{where } V \ll kA^2 \text{ and } \omega = \omega_0 \tag{8.12}$$

Arvind Raman also provides [1] the 'linear equivalent' expression for $\tan \phi$

$$\tan\,(\phi) = \frac{\left(\frac{\omega}{\omega_0 Q'}\right)}{\left(\frac{{\omega_0'}^2 - \omega^2}{\omega_0^2}\right)} \tag{8.13}$$

Again, equation (8.13) is simplified at $\omega = \omega_0$

$$\tan\,(\phi) = \frac{\frac{1}{Q'}}{\left[\frac{\omega_0'}{\omega_0}\right]^2 - 1} \tag{8.14}$$

Combining equations (8.7), (8.8) and (8.14)

$$\tan\,(\phi) = \frac{\frac{1}{Q} + \frac{E_{dis}}{\pi kA^2}}{\left[1 - \frac{V}{kA^2}\right]^2 - 1} \quad \text{where } \frac{\omega_0'}{\omega_0} = 1 - \frac{V}{kA^2} \tag{8.15}$$

Equation (8.15) can be simplified for the weakly perturbed oscillator as follows:

$$\tan\,(\phi) = \frac{\frac{1}{Q} + \frac{E_{dis}}{\pi kA^2}}{\left[1 + \left[\frac{V}{kA^2}\right]^2 - 2\frac{V}{kA^2}\right] - 1}$$

$$\tan\,(\phi) \approx \frac{\frac{1}{Q} + \frac{E_{dis}}{\pi kA^2}}{-2\frac{V}{kA^2}} \quad \text{where } V \ll kA^2 \text{ and } \omega = \omega_0 \tag{8.16}$$

to obtain the term $\tan \phi$ where $E_{dis} = 0$ is relevant, since this allows determining the frequency shift (see equations (8.1) and (8.2)). Then,

$$\tan\,(\phi) \approx \frac{1}{-2Q\frac{V}{kA^2}} \quad \text{where } V = -\frac{1}{2}F_0 A \cos \phi \text{ and } F_0 = \frac{kA_0}{Q} \tag{8.17}$$

Finally, simplifying equation (8.17)

$$\tan(\phi) \approx \frac{A}{A_0 \cos\phi} \tag{8.18}$$

8.1.2 The transfer function and amplitude decay

The general transfer function results can be written in terms of V and E_{dis} by combining equations (8.6)–(8.8)

$$\frac{kA}{F_0} = \frac{1}{\sqrt{\left(\left[1 - \frac{V}{kA^2}\right]^2 - \frac{\omega^2}{\omega_0^2}\right)^2 + \left(\frac{\omega}{\omega_0}\left[\frac{1}{Q} + \frac{\omega_0 E_{dis}}{\omega \pi k A^2}\right]\right)^2}}$$

$$\frac{kA}{F_0} \approx \frac{1}{\sqrt{\left(1 - 2\frac{V}{kA^2} - \frac{\omega^2}{\omega_0^2}\right)^2 + \left(\frac{\omega}{\omega_0}\left[\frac{1}{Q} + \frac{\omega_0 E_{dis}}{\omega \pi k A^2}\right]\right)^2}} \quad V \ll kA^2 \tag{8.19}$$

Equation (8.19) says that the virial V and E_{dis} control the amplitude decay together with the Q factor and the ratio $\frac{\omega}{\omega_0}$. For example, if $\frac{\omega}{\omega_0} = 1$, and $V - E_{dis} = 0$, equation (8.19) reduces to the well-known expression for F_0 at $\frac{\omega}{\omega_0} = 1$ in the linear case

$$\frac{kA}{F_0} = \frac{1}{\frac{1}{Q}}$$

$$F_0 = \frac{kA}{Q} \tag{8.20}$$

There is a case that was not discussed in the linear theory and is useful in the analysis of amplitude decay in the nonlinear theory. This is the case where there is drive but there is no viscosity, namely

$$m\ddot{z} + kz = F_0 \cos \omega t \tag{8.21}$$

As an exercise, the reader can use the methods detailed in the previous chapters to find that, in the steady state, the solution to equation (8.21) is

$$z = A\cos(\omega t) \text{ where } A = \frac{F_0}{m[\omega_0^2 - \omega^2]} \tag{8.22}$$

According to (8.22), the amplitude A will be maximum when $\omega = \omega_0$. The drive will pump energy into the system at the frequency ω. Note that there is no mechanism for dissipation in equation (8.21). The first implication is that $\omega_r = \omega_0$. That is, if there is no mechanism for dissipation the resonant frequency is identical to the natural frequency. Second, at $\omega = \omega_0$ the amplitude A tends to infinity with time since, as energy is delivered into the system, it all adds to A. A third implication is that, even in the absence of dissipation, the amplitude A will be finite outside $\omega = \omega_0$. This is because, in such cases, the drive eventually adds nothing to the oscillation. This can happen when the power delivered by the drive with time tends to zero. Practically, such phenomenon can be experienced when pushing someone in a swing and the pushing happens at the 'wrong' time or the 'wrong' frequency. Pushing in such cases damps the oscillation. More on equation (8.21) can be found in chapter 21 of Feynman's lectures on physics.

San Paulo and Garcia [2] explored the amplitude decay in AM AFM (nonlinear case) as early as 2001. From the discussion of equations (8.21) and (8.22) it is clear that variations in the effective value of ω_0, i.e., ω_0', can change A. In chapters 2 and 4, it was also shown that dissipation adds to amplitude decay in the linear case. In summary, the linear theory teaches us that the amplitude A is controlled by ω_0 and energy dissipation, i.e., Q (see chapter 4 and figures 4.4–4.6).

Arguably, when invoking the virial theorem, San Paulo and Garcia had in mind finding the mechanism for amplitude decay in AM AFM. They also indirectly found the relationship between AM and FM AFM since they found a relationship between the virial V and the cosine of the angle and recalled that Giessibl had already found in 1997 that the virial is proportional to a frequency (natural) shift. In short, the authors found that the expression controlling E_{dis} was related to the sine of the phase shift while the virial V was related to the cosine (equations (7.61) and (7.62)). These two expressions in combination with Pythagoras (equation (8.5)) theorem can be used to find the relationship between A, E_{dis} and V as follows

$$1 = \sin^2 \phi + \cos^2 \phi$$

The cosine and the sine of the phase shift are given by the expressions containing the virial V and E_{dis}, respectively

$$\cos \phi = \frac{A}{F_0}[k - \omega^2 m] - \frac{2}{AF_0}V$$

$$\sin \phi = \frac{1}{\pi AF_0}E_{dis} + \frac{m\omega_0}{QF_0}\omega A$$

Combining equations (7.61), (7.62) and (8.5)

$$1 = \left[\frac{1}{\pi AF_0}E_{dis} + \frac{m\omega_0}{QF_0}\omega A\right]^2 + \left[\frac{A}{F_0}[k - \omega^2 m] - \frac{2}{AF_0}V\right]^2 \tag{8.23}$$

The drive in equation (8.23) can be factorized in terms of F_0. From F_0 an amplitude A_0 can be defined, i.e., the unperturbed amplitude, which differs from the perturbed amplitude A. It is useful to define the drive in terms of A_0 because it can be experimentally measured. In dynamic AFM A_0 is the value of oscillation amplitude A obtained in the absence of tip-sample interaction. Then, the system behaves like the driven oscillator. Thus, A_0 can be identified with the amplitude A obtained when

$E_{dis} = V = 0$, i.e., for the conditions of the linear theory or driven oscillator. Then, equation (8.23) reduces to

$$\frac{F_0^2}{A^2} = [k - \omega^2 m]^2 + \left[\frac{m\omega_0}{Q}\omega\right]^2$$

$$F_0^2 = A^2\left[[k - \omega^2 m]^2 + \left[\frac{m\omega_0}{Q}\omega\right]^2\right] \tag{8.24}$$

Writing equation (8.24) terms of A

$$A^2 = \frac{F_0^2}{[k - \omega^2 m]^2 + \left[\frac{m\omega_0}{Q}\omega\right]^2}$$

$$A = \frac{F_0}{\sqrt{[k - \omega^2 m]^2 + \left[\frac{m\omega_0}{Q}\omega\right]^2}} \tag{8.25}$$

Or equivalently,

$$A = \frac{F_0}{\sqrt{m^2(\omega_0^2 - \omega^2)^2 + b^2\omega^2}}$$

$$F_0 = A\sqrt{m^2(\omega_0^2 - \omega^2)^2 + b^2\omega^2}$$

The free amplitude A_0 and the drive force F_0 are simultaneously defined from equation (8.9), then (A_0 defined)

$$A_0 = \frac{F_0}{\sqrt{m^2(\omega_0^2 - \omega^2)^2 + b^2\omega^2}} \tag{8.26}$$

The expression above is the standard result that relates the amplitude with the drive (equation (8.9)) but it is useful to define A_0 as shown in equation (8.26) since it is a way to distinguish the unperturbed amplitude $A_0\,(E_{dis} = V = 0)$ with the perturbed amplitude, i.e., nonlinear.

Having clarified the difference between A_0 and A, the relation between F_0 and A_0, and that the standard expression relating A and F_0 can be recovered from the nonlinear expression in equation (8.23) (equations (8.24)–(8.26) and equation (2.8)), the objective is to find the roots of A in the nonlinear expression (equation (8.23)) and to express them in terms of E_{dis} and V.

First, F_0 can be factorized from equation (8.23)

$$F_0^2 = \left[\frac{1}{\pi A}E_{dis} + \frac{m\omega_0}{Q}\omega A\right]^2 + \left[A[k - \omega^2 m] - \frac{2}{A}V\right]^2 \tag{8.27}$$

Second, equation (8.27) can be factorized as follows

$$F_0^2 = \left[\frac{1}{\pi A}E_{dis}\right]^2 + \left[\frac{m\omega_0}{Q}\omega A\right]^2 + 2\left[\frac{1}{\pi A}E_{dis}\frac{m\omega_0}{Q}\omega A\right] \\ + [A[k-\omega^2 m]]^2 + \left[\frac{2}{A}V\right]^2 - 2\left[A[k-\omega^2 m]\frac{2}{A}V\right] \tag{8.28}$$

Factorizing in terms of A and simplifying

$$F_0^2 = \frac{1}{A^2}\left[\frac{1}{\pi}E_{dis}\right]^2 + A^2\left[\frac{m\omega_0}{Q}\omega\right]^2 + \left[\frac{2}{\pi}E_{dis}\frac{m\omega_0}{Q}\omega\right] \\ + A^2[k-\omega^2 m]^2 + \frac{1}{A^2}[2V]^2 - [[k-\omega^2 m]4V] \tag{8.29}$$

Multiplying all terms by A^2 and collecting terms

$$A^4\left[\left[\frac{m\omega_0}{Q}\omega\right]^2 + [k-\omega^2 m]^2\right] + A^2\left[\frac{2}{\pi}E_{dis}\frac{m\omega_0}{Q}\omega - 4[k-\omega^2 m]V - F_0^2\right] \\ + \left[\left[\frac{1}{\pi}E_{dis}\right]^2 + [2V]^2\right] = 0 \tag{8.30}$$

The expression in equation (8.30) is a quadratic equation in terms of A^2. This is true because the term A_0 is not the same as the perturbed amplitude A. Note that while equation (8.30) produces an explicit solution for A for arbitrary values of ω, F_0, E_{dis}, V, Q and ω_0, equation (8.19) does not. An explicit solution for A was found by Álvarez Amo [3] (equations (2.20)–(2.25) in the thesis) assuming that $F_0 = \frac{kA_0}{Q}$. This is approximately true provided $\omega \approx \omega_0$. The advantage of equation (8.30) is that there is no assumption about F_0, or, equivalently, it is valid for any drive frequency ω and drive force F_0.

Simplifying equation (8.30) by driving at $\omega = \omega_0$

$$A^4 + A^2\left[\frac{2Q}{\pi k}E_{dis} - \left[\frac{QF_0}{k}\right]^2\right] + \left[\left[\frac{Q}{\pi k}E_{dis}\right]^2 + \left[\frac{2Q}{k}V\right]^2\right] = 0 \tag{8.31}$$

Equation (8.30) is general for any ω and does not assume $F_0 = \frac{kA_0}{Q}$. Nevertheless, in equation (8.31) the expression $F_0 = \frac{kA_0}{Q}$ is valid since $\omega = \omega_0$. Then, equation (8.31) can be further reduced to give

$$A^4 + A^2\left[\frac{2Q}{\pi k}E_{dis} - A_0^2\right] + \left[\left[\frac{Q}{\pi k}E_{dis}\right]^2 + \left[\frac{2Q}{k}V\right]^2\right] = 0 \quad \omega = \omega_0 \tag{8.32}$$

The expression in equation (8.32) is equivalent to the result of Álvarez Amo [3] in his equation (2.26) and the solution was first reported in 2001 by San Paulo and García [2] in their equation (9). The solution can be easily found using software like Matlab. A script to solve equation (8.32) is

```
Matlab script to solve (8.32)
syms A a b c
F=A^4 + A^2*(2*a–b) + a^2+c;
Sa = solve(F, A)
```

The expression in equation (8.32) has been simplified in order to solve it in Matlab by defining

$$a = \frac{Q}{\pi k}E_{dis}$$

$$b = A_0^2$$

$$c = \left[\frac{2Q}{k}V\right]^2 \tag{8.33}$$

Of the four solutions only two are valid since, from energy conservation, $A \leqslant A_0$. The two valid solutions are

$$A = \left[\frac{b}{2} - a \pm \frac{1}{2}\sqrt{b^2 - 4ab - 4c}\right]^{1/2} \tag{8.34}$$

Combining equations (8.33) and (8.34) and arranging the terms

$$= \frac{A_0}{\sqrt{2}}\left[1 - \frac{2Q}{\pi k A_0^2}E_{dis} \pm \sqrt{1 - \frac{4\frac{QA_0^2}{\pi k}E_{dis} - 4\left[\frac{2Q}{k}V\right]^2}{A_0^4}}\right]^{1/2} \tag{8.35}$$

Equation (8.35) is valid for interactions with dissipation and with conservative terms, i.e., any nonlinear interaction. The equation shows that both E_{dis} and V contribute to the amplitude decay. If we allow $E_{dis} = V = 0$, equation (8.35) is undetermined, i.e., $A = 0$ is the only solution. But San Paulo and García [2] showed by numerical integration of the equation of motion that equation (8.35) is valid for any interaction provided $\omega = \omega_0$.

Equation (8.32) can be written for the case for conservative interactions only, i.e., $E_{dis} = 0$. The expression shows that A depends on A_0, V, Q and $k\,(\omega = \omega_0)$

$$A^4 - A^2A_0^2 + \left[\frac{2Q}{k}V\right]^2 = 0 \text{ where } E_{dis} = 0 \text{ and } \omega = \omega_0 \tag{8.36}$$

```
Matlab script to solve (8.36)
syms theta a b
F=A^4 -(A^2)a + b;
Sa = solve(F,A)
```

The two valid solutions of equation (8.36) are given by the script above.

$$= \frac{A_0}{\sqrt{2}}\left[1 \pm \sqrt{1 - \frac{16Q^2}{k^2A_0^4}V^2}\right]^{1/2} \quad \text{where } E_{dis} = 0 \text{ and } \omega = \omega_0 \tag{8.37}$$

The expression in equation (8.37) is equivalent to equation (11) presented in 2001 by San Paulo and García [2]. It can also be found as equation (2.28) in the thesis of Álvarez Amo [3]. In their paper, San Paulo and García solve the equation of motion numerically to show that equation (8.34) is exact when only conservative forces are present in the nonlinear term. Equation (8.37) can be combined with equation (6.47) to express the amplitude decay in term of the cosine of the phase shift

$$A = \frac{A_0}{\sqrt{2}}\left[1 \pm \sqrt{1 - \frac{16Q^2}{k^2A_0^4}\left[\frac{1}{2}F_0 A\cos\phi\right]^2}\right]^{1/2} \quad \text{where } V = -\frac{1}{2}F_0 A\cos\phi \text{ and } F_0 = \frac{kA_0}{Q} \tag{8.38}$$

Simplifying

$$A = \frac{A_0}{\sqrt{2}}\left[1 \pm \sqrt{1 - 4\frac{A^2}{A_0^2}\cos^2(\phi)}\right]^{1/2} \quad \text{where } E_{dis} = 0 \text{ and } \omega = \omega_0 \tag{8.39}$$

Equation (8.39) shows that by expressing V in terms of F_0 and the cosine of the phase shift ϕ (6.47) the equations expressing the amplitude decay A are shown to express implicit relations. The same is true for E_{dis} since E_{dis} depends on A. On the other hand, one can argue that E_{dis} and V can be taken as experimentally computable quantities directly extracted from physical phenomena and inputted into expressions such as that in equation (8.39).

Finally, factorizing A in equation (8.39) produces a true expression for the amplitude decay without implicit relationships with A. The steps for factorization are shown below

$$\frac{2A^2}{A_0^2} = 1 - \sqrt{1 - 4\frac{A^2}{A_0^2}\cos^2(\phi)}$$

$$\left[\frac{2A^2}{A_0^2} - 1\right]^2 = 1 - 4\frac{A^2}{A_0^2}\cos^2(\phi)$$

$$\left[\frac{2A^2}{A_0^2}\right]^2 - \frac{4A^2}{A_0^2} + 1 = 1 - 4\frac{A^2}{A_0^2}\cos^2(\phi)$$

$$\frac{4}{A_0^4}A^4 + \frac{4}{A_0^2}[\cos^2(\phi) - 1]A^2 = 0$$

$$\frac{4}{A_0^2}\left[\frac{1}{A_0^2}A^4 + [\cos^2(\phi) - 1]A^2\right] = 0$$

$$\frac{1}{A_0^2}A^2 + \cos^2(\phi) - 1 = 0 \quad \text{where} \cos^2(\phi) - 1 = \sin^2(\phi)$$

$$\frac{A^2}{A_0^2} = -\sin^2(\phi)$$

$$A = A_0 \sin(\phi) \quad \text{where } E_{dis} = 0 \text{ and } \omega = \omega_0 \tag{8.40}$$

Equation (8.40) shows that when only the virial V, i.e., a conservative force, is responsible for the decay of the perturbed amplitude A, i.e., $A \leqslant A_0$, the amplitude A can be found directly from A_0 and the sine of the phase shift alone. This is consistent with equation (6.31) and the relations shown in figure 6.6. In summary, setting $E_{dis} = 0$ provides an expression for the amplitude decay A when $E_{dis} = 0$ and $\omega = \omega_0$.

8.1.3 The limit of small amplitudes

If the amplitude of oscillation A is small compared to the variations of the nonlinear term, the linear approximation can be exploited. It is important to note that the unperturbed, or linear, amplitude A_0 can take on any value, that is, the derivation involves A not the drive F_0.

When the oscillation A is small compared to the variations in the nonlinear term F_{ts} (see figure 7.1), the curve of F_{ts} can be approximated to a straight line. This is why transforming F_{ts} to a straight line is termed 'linearizing' the equation.

The tangent to a curve in a point gives the slope of the curve at that point. For an arbitrary point at d_0, the tangent of F_{ts} is

$$\frac{\partial F_{ts}}{\partial d}(d_0) = -k_{ts}(d_0) \tag{8.41}$$

where k_{ts} is the tangent or slope at a point d_0. If the value of k_{ts} is approximately constant in a neighbourhood of points around d_0, i.e., $d_0 - A < d < d_0 + A$, the nonlinear term F_{ts} can be approximated as

$$\frac{\partial F_{ts}}{\partial d}(d) \approx \frac{\partial F_{ts}}{\partial d}(d_0) \quad for\, d_0 - A \leqslant d \leqslant d_0 + A \tag{8.42}$$

$$F_{ts}(d) \approx -k_{ts}(d_0)A\cos(\omega t - \phi) \tag{8.43}$$

where the negative sign in (8.43) implies that the gradient should be positive for repulsive forces and negative for attractive forces. This sign is thus related the physics of the phenomena under consideration. In AFM it must be negative. Figure 2.1 has been reproduced here as figure 8.2 to support the exposition since the geometry of the problem can be used to derive the expressions.

Only the dynamic part of z is used here since, as shown in chapter 6 when deriving the virial, the equilibrium position z_0 does not add to the virial V.

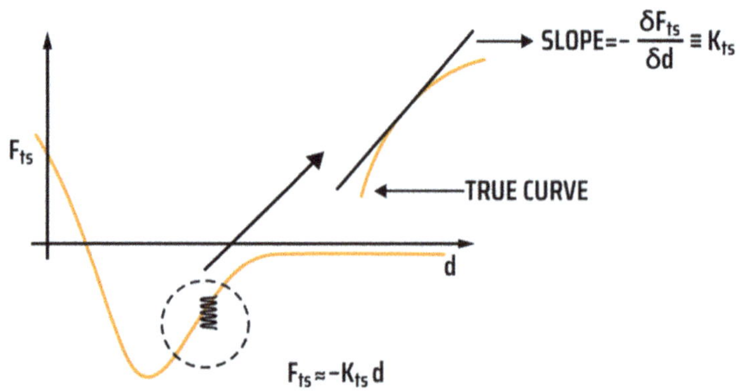

Figure 8.1. Illustrations depicting of nonlinear force F_{ts} in terms of distance d where, as indicated, if the oscillation amplitude A is small enough, the equation of motion can be linearized. The slope, gradient or tangent at the position of equilibrium z_0.

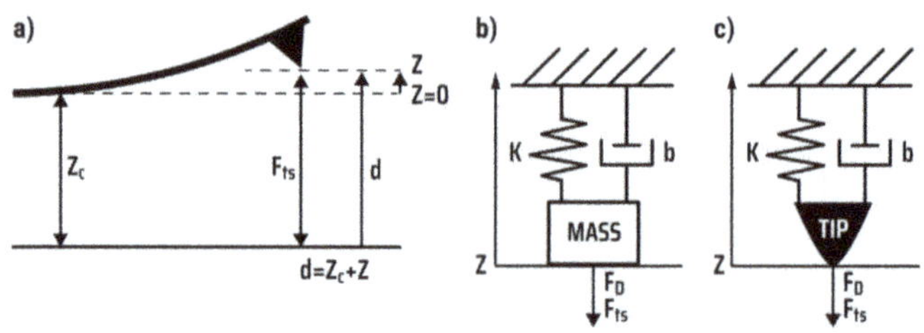

Figure 8.2. (a) Schematic of an AFM cantilever from which geometrical constraints can be derived. (b) and (c) Rheological models of the tip' motion. Both models are mathematically equivalent but the illustration in (c) showcases that the motion is discussed in terms of the tip.

In summary, d_0 is identified with the equilibrium position of oscillation, that is, with the neutral position of equilibrium, i.e., $z = z_0$ or mean deflection. The mean deflection is given by the average force per cycle and there is oscillation around this neutral position. For this reason, in the limit of small oscillations, the tangent, or derivative of F_{ts} is taken in the neutral position of oscillation and assumed to be approximately constant for the range of d covered (equation (8.42)). The linear approximation is thus constrained by equation (8.42), i.e., the better equation (8.42) holds the better the linear approximation. Physically, the tangent of the curve depends on the physical phenomena but the AFM user can improve the approximation by reducing the oscillation amplitude A (see figure 8.1).

The general expression for the virial V is

$$V \equiv \langle Fz \rangle = \frac{1}{T}\int_0^T Fz dt \tag{8.44}$$

where

$$z = A\cos(\omega t - \phi)$$

The virial V for a force based on the derivative or slope at d can be written by combining equations (8.42) and (8.43)

$$V = -\frac{1}{T}\int_0^T k_{ts}(d)z^2 dt$$

In the limit of small oscillations A the slope (also termed gradient or tangent) is approximately constant (see figure 7.1), then

$$V \approx -\frac{1}{T}k_{ts}(d_0)\int_0^T z^2 dt \quad \text{where } \frac{\partial F_{ts}}{\partial d}(d) \approx \frac{\partial F_{ts}(d)}{\partial d}(d_0) \tag{8.45}$$

A change of variable is then applied as shown below

$$z = A\cos\varphi \quad \text{where } \varphi = \omega t - \phi \text{ and } d\varphi = \omega dt$$

Combining (8.44) and (8.45)

$$V \approx -\frac{1}{2\pi}k_{ts}(d_0)A^2\int_{\varphi=0}^{\varphi=2\pi}\cos^2(\varphi)d\varphi \tag{8.46}$$

Since the integral in equation (8.46) is π, the virial for small oscillation amplitudes is

$$V \approx -\frac{1}{2}k_{ts}A^2 \quad \frac{\partial F_{ts}}{\partial d}(d) \approx \frac{\partial F_{ts}(d)}{\partial d}(d_0) \equiv k_{ts} \tag{8.47}$$

where k_{ts} is the effective spring constant resulting from the nonlinear term F_{ts} and computed as its gradient.

Combining equation (8.47) with equation (6.48) gives the frequency shift for a weakly perturbed oscillator and the small amplitude, i.e., linear, approximation. Then

$$\frac{\Delta\omega_0}{\omega_0} = -\frac{1}{kA^2}V$$

$$\frac{\Delta\omega_0}{\omega_0} \approx -\frac{1}{k}\left[\frac{1}{2}k_{ts}A^2\right]$$

$$\frac{\Delta\omega_0}{\omega_0} \approx -\frac{1}{2}\frac{k_{ts}}{k} \quad \text{where } k_{ts} \equiv \frac{\partial F_{ts}}{\partial d}(d) \tag{8.48}$$

In equation (8.48) the sign is negative because the negative in the original equation already contained the negative of the gradient. The result in equation (8.48) was already reported [4] in 1991 one FM AFM was first introduced. Giessibl also acknowledged [5] that his equation (6.48) gave the right solution in the limit of small amplitudes.

The frequency shift for the small amplitude A limit can also be computed directly via the linear equation of motion (equation (5.2)). A main result from the linear theory is that m, k and ω_0 are related

$$\omega_0^2 = \frac{k}{m}$$

For small perturbations, i.e., $\Delta\omega_0 \ll \omega_0$

$$[\omega_0 + \Delta\omega_0]^2 = \frac{k'}{m} \Delta\omega_0 \ll \omega_0 \quad \text{and} \quad k' = k - k_{ts} \tag{8.49}$$

where k' is the effective spring constant and where k_{ts} has the minus sign to indicate that repulsive forces produce negative sloped. Factorizing

$$\omega_0^2 + \Delta\omega_0^2 + 2\omega_0\Delta\omega_0 = \frac{k - k_{ts}}{m}$$

The simplification for weakly perturbed oscillators, i.e., $\Delta\omega_0 \ll \omega_0$, produces

$$\omega_0^2 + 2\omega_0\Delta\omega_0 \approx \frac{k}{m} - \frac{k_{ts}}{m} \tag{8.50}$$

Finally, simplifying equation (8.50)

$$\omega_0\Delta\omega_0 m \approx -\frac{1}{2}k_{ts}$$

$$\omega_0\Delta\omega_0\left[\frac{k}{\omega_0^2}\right] \approx -\frac{1}{2}k_{ts}$$

$$\frac{\Delta\omega_0}{\omega_0} \approx -\frac{1}{2}\frac{k_{ts}}{k} \tag{8.51}$$

Equation (8.51) coincides with equation (8.48) and it is thus in agreement with the weak perturbation theory presented here.

For large amplitudes, other than the problem of linearity, i.e., the gradient is not constant during the oscillations, higher harmonics are excited and cannot be neglected [6–8]. Oscillating in highly damped medium might also distort the harmonicity of the motion [9–11]. In short, the small amplitude approximation requires that A is much smaller than the features being probed. The attractive forces probed in AFM are ~1 nm [12]. Thus the amplitude to probe nanoscale forces much be $A \ll 1$ nm. This is a typical characteristic of microscopy.

8.1.4 Thermal energy, energy dissipation and virial

The smallest virial V that can be detected in an oscillator above thermal noise can be computed from the equipartition theorem [13]

$$\frac{1}{2}k\langle \dot{z}^2\rangle = \frac{1}{2}k_B T \ \omega = \omega_0 \tag{8.52}$$

where k_B is Boltzmann's constant and T is the temperature in Kelvin.

From equation (6.35)

$$\langle KE\rangle = \frac{1}{2}m\langle \dot{z}^2\rangle$$

$$\langle KE\rangle \frac{1}{2}E_T = \frac{1}{2}\left[\frac{1}{2}kA^2\right]$$

Combining equations (8.52) and (6.35) provides the condition for the minimum amplitude A to be detected at temperature T (in Kelvin) is

$$\frac{1}{4}kA^2 = \frac{1}{2}k_B T$$

$$A^2 = 2\frac{k_B}{k}T \tag{8.53}$$

From equation (8.35), the amplitude A is found for any energy dissipation E_{dis} and virial V at $\omega = \omega_0$. Combining equation (8.53) and equation (8.35)

$$\frac{A_0^2}{2}\left[1 - \frac{2Q}{\pi k A_0^2}E_{dis} \pm \sqrt{1 - \frac{4\frac{QA_0^2}{\pi k}E_{dis} - 4\left[\frac{2Q}{k}V\right]^2}{A_0^4}}\right] = 2\frac{k_B}{k}T \quad \omega = \omega_0 \tag{8.54}$$

The minimum detectable E_{dis} and V follow from equation (8.54). We leave this problem unresolved here but suffice it to say that it requires that E_{dis} and V are explicitly written in term of T. The equation can be simplified if either E_{dis} or V are zero. For example, if $E_{dis} = 0$, the amplitude A in terms of V is simplified to equation (8.37). Combining equations (8.37) and (8.53)

$$\frac{A_0^2}{2}\left[1 \pm \sqrt{1 - \frac{16Q^2}{k^2A_0^4}V^2}\right] = 2\frac{k_B}{k}T \quad \text{where } E_{dis} = 0 \text{ and } \omega = \omega_0 \tag{8.55}$$

$$\sqrt{1 - \frac{16Q^2}{k^2A_0^4}V^2} = 4\frac{1}{A_0^2}\frac{k_B}{k}T - 1$$

$$1 - \frac{16Q^2}{k^2A_0^4}V^2 = \left[4\frac{1}{A_0^2}\frac{k_B}{k}T - 1\right]^2$$

$$V^2 = \frac{k^2A_0^4}{16Q^2}\left[1 - \left[4\frac{1}{A_0^2}\frac{k_B}{k}T - 1\right]^2\right] \text{minimum } V \text{ at } T \text{ where } E_{dis} = 0 \text{ and } \omega = \omega_0 \tag{8.56}$$

From equation (8.56) it follows that when $T = 1$, the minimum detectable virial V is zero as required. This assumes however that no other sources of noise, i.e., electric, etc, are present. For this reason equation (8.56) is a limiting case, i.e., the physically minimum detectable V at temperature T. A similar relationship can be found for the phase shift from equation (8.40) in combination with equation (8.53). We also leave the problem unresolved.

8.2 Bimodal

In the first chapter, the first equation discussed in the analysis of oscillations was that of a cantilever parametrized in terms of beam theory [14] (equation (1) in chapter 2 on linear theory)

$$EI\frac{\partial}{\partial x^4}\left[w(x,\, t) + a_1\frac{\partial w(x,t)}{\partial t}\right] + \rho bh\frac{\partial^2 w(x,t)}{\partial t^2}$$
$$= -a_0\frac{\partial w(x,t)}{\partial t} + \delta(x - L)[F_{\text{exc}}(t) + F_{ts}(d)]$$

where all the parameters are defined there. Equations (2.2)–(2.3) showed that equation (2.1) can be reduced to a set of equations that represent the vertical motion of the beam at $x = L$, i.e., at the edge of the cantilever (see the schematic in figure 8.1). In short, equation (2.1) is equivalent to a set of anharmonic differential equations [15, 16] (equation (2.3) on linear theory)

$$\ddot{z}_m(t) + \frac{\omega_m}{Q_m}\dot{z}_m(t) + \omega_m^2 z_m(t) = \frac{F_{\text{exc}}(t) + F_{ts}(d)}{m},\; m(\text{subscript}) = 1,\, 2,\, \ldots$$

where $z(t)$ is the modal projection of the tip motion (figure 8.1). All the other parameters are defined in chapter 2 when discussing equations (2.1)–(2.3). In equation (2.3) all the parameters are equivalent to those appearing in the equation of motion analysed in section 2 so far (equation (5.3)) with the only difference being the appearance of a subscript m. This subscript stands for mode m. The subscript m should not be confused with the mass m. Using m rather than n for mode is useful because Fourier series are typically developed using n for harmonic. It is now more or less standard terminology in AFM to use n for harmonics and m [17] or i [18] for mode.

While all modes have a finite response at all frequencies ω, the signal for each mode m is significant mostly near the resonance of each mode m. The response of the first two modes is illustrated in figure 2.5—reproduced here as figure 8.3.

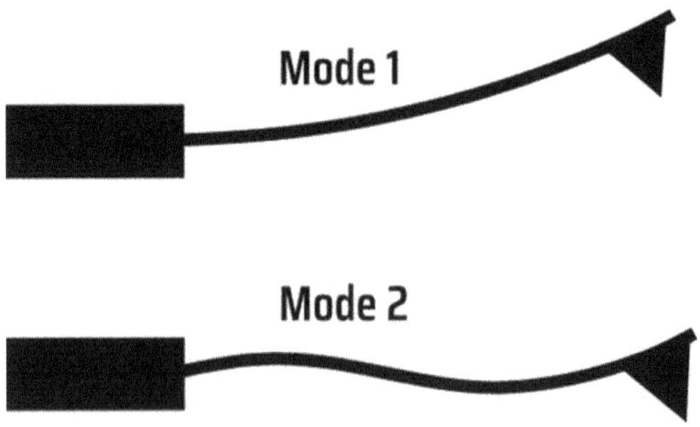

Figure 8.3. Illustration of the first two modes of a cantilever. See literature for details [19]. Reprinted from [19], with the permission of AIP Publishing.

In short, the equations of motion for the first two modes are

$$\ddot{z}_1 = -k_1 z_1 - \frac{m\omega_{01}}{Q_1}\dot{z}_1 + F_{01}\cos\omega_1 t + F_{02}\cos\omega_2 t + F_{ts}(z_1 + z_2) \quad (8.57)$$

$$m\ddot{z}_2 = -k_2 z_2 - \frac{m\omega_{02}}{Q_2}\dot{z}_2 + F_{01}\cos\omega_1 t + F_{02}\cos\omega_2 t + F_{ts}(z_1 + z_2) \quad (8.58)$$

where the subscripts 1 and 2 stand for mode 1 and 2, respectively. The subscripts 01 and 02 stand for natural frequency of mode 1 and 2, respectively. In terms of the drive force they stand for the drive F_0 at mode 1 and 2, respectively. Here higher modes will be ignored in order to develop a theory based on the virial V and the energy dissipation E_{dis} for each mode. This theory is will be briefly discussed in the remainder of this book [19, 20]. Lozano and García provided a general theory in 2008 [18] in terms of the virial theorem and supplemented their results and interpretation via numerical integration of the equations of motion. Experimental 'amplitude' (power density) versus frequency ω curves, are shown in figure 2.4—reproduced here as figure 8.4.

Key phenomena from equations (8.57) and (8.58) relate to the presence of two drives, one exciting at or near the natural frequency of mode 1 ($m = 1$) ω_{01}, and the other at or near the natural frequency of mode 2 ($m = 2$) ω_{01}. The position z is further expressed as the sum of z_1 and z_2

$$\begin{aligned} z(t) &= z_1(t) + z_2(t) + O(\varepsilon) \\ &\approx A_1\cos\left(\omega_1 t - \phi_1\right) + A_2\cos\left(\omega_2 t - \phi_2\right) \end{aligned} \quad (8.59)$$

where $O(\varepsilon)$ is the term carrying the contributions of higher harmonics and higher modes, i.e., it is the error not accounted for by the first two terms. The force F_{ts}, i.e., the nonlinear term, is also, and approximately, determined by z_1 and z_2

$$F_{ts}(z) \approx F_{ts}(z_1 + z_2) \quad (8.60)$$

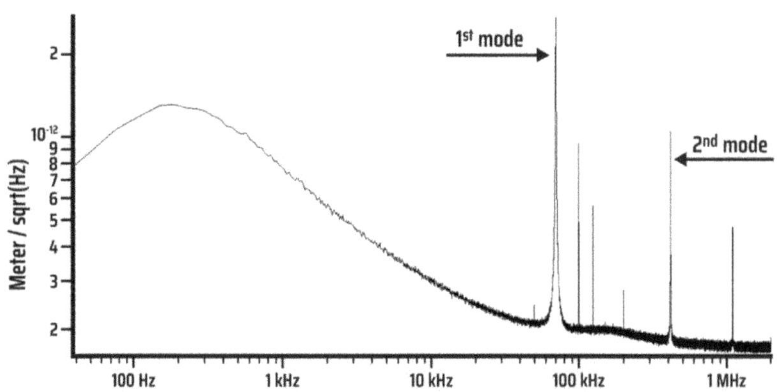

Figure 8.4. Standard frequency sweep for an AFM cantilever showing the response of modes 1 and 2. The response of the 3rd mode can also be seen near 1MHz. The other peaks are noise, i.e., electronic or other.

we emphasize $z_1(t)$ and $z_2(t)$ are the positions of modes 1 and two. Another approximation can be applied here by assuming that the responses for each mode coincide with the frequencies of excitation at each mode. This is a good approximation if the first drive acts at a frequency near the natural frequency of the first mode, and the second near that of the second mode. Then the subscripts in equations (8.59) and (8.60) can be taken to represent the harmonics at the drive frequencies ω_1 and ω_1, respectively. It turns out that the virial V and energy dissipation expressions for each mode are identical to those already discussed. This is because the equations of motion (equations (8.57)–(8.58)) are identical to equation (3) (provided the approximations in equation (8.59) are assumed and the interpretation given for z is considered, namely, each drive only excites one mode because each drive drives near the resonance of each mode). The virials V_1 and V_2 stand for the virials of mode 1 and 2, respectively, and can be written as

$$V_1 = \langle F_{ts} z_1 \rangle = \frac{1}{T}\int_0^T F_{ts} z_1 dt$$

$$V_1 = -\frac{1}{2}F_{01}A_1 \cos\phi_1 + \frac{1}{2}A_1^2[k_1 - \omega_{01}^2 m] \text{ an arbitrary } \omega$$

$$V_1 = -\frac{1}{2}F_{01}A_1 \cos\phi_1 \quad \omega = \omega_0 \tag{8.61}$$

$$V_2 = \langle F_{ts} z_2 \rangle = \frac{1}{T}\int_0^T F_{ts} z_2 dt$$

$$V_2 = -\frac{1}{2}F_{02}A_2 \cos\phi_2 + \frac{1}{2}A_2^2[k_2 - \omega_{02}^2 m] \text{ an arbitrary } \omega$$

$$V_2 = -\frac{1}{2}F_{02}A_2 \cos\phi_2 \quad \omega = \omega_0 \tag{8.62}$$

Note that for both modes the fundamental period T is used to derive the expressions. This means that V_2 is an average for a fundamental cycle of oscillation. For the energy dissipation we have (from equation (6.26)) [19]

$$E_{dis(1)} = -\int_0^T F_{ts}\dot{z}_1 dt$$

$$E_{dis(1)} = \frac{\pi k_1 A_{01} A_1}{Q_1}\left[\sin\phi_1 - \frac{A_1}{A_{01}}\right] \quad \omega_1 = \omega_{01} \tag{8.63}$$

$$E_{dis(2)} = -\int_0^T F_{ts}\dot{z}_2 dt$$

$$E_{dis(2)} = \pi F_{02}A_2 \sin\phi_2 - \frac{m\omega_{02}}{Q_2}\omega_2 \pi A_2^2 \text{ an arbitrary } \omega$$

$$E_{dis(2)} = \frac{\pi k_2 A_{02} A_2}{Q_2}\left[\sin\phi_2 - \frac{A_2}{A_{02}}\right] \quad \omega_2 = \omega_{02} \tag{8.64}$$

As a note, it should be emphasized that it is clear that the drives will affect both modes. In the expressions above the approximation is that the displacement z near the drives ω_1 and ω_1 can be accounted for by the response at each mode only. The numerical integration of the equation of motion however shows that this is a good approximation, particularly in air and vacuum where the Q factor is high [15, 17–19].

8.2.1 The frequency shift of the second mode

For the frequency shift of the second mode, the derivation does not fully follow from the analysis of the first mode. The analysis of Giessibl however still holds for the higher modes. Then,

$$\frac{\Delta f_{02}}{f_{02}} \approx -\frac{1}{k_2 A_2^2}\langle F_{ts} z_2 \rangle \tag{8.65}$$

It turns out that solving equation (8.65) is very difficult because typically $A_1 \gg A_2$. Instead of solving it directly, Kawai *et al* [21] proposed that the general linear result in equations (8.48) or (8.58) apply to the second mode, provided the oscillation of the second mode A_2 is much smaller than that of the first mode A_1 and that the time averaged derivative of F_{ts} is used instead of the derivative at z_0. The proposed approximation is

$$k_2 \frac{\Delta f_2}{f_2} \cong -\frac{1}{4\pi} \int_0^{2\pi} \frac{\partial F_{ts}(d)}{\partial d}(z_0 + A_1 \cos\theta_1) d\theta_1 \quad A_1 \gg A_2 \tag{8.66}$$

The above expressions can be combined as follows:

$$\frac{1}{T}\oint \frac{\partial F_{ts}}{\partial \mathrm{d}} dt \approx -2k_2 \frac{\Delta f_{02}}{f_{02}} \tag{8.67}$$

$$V_2 \approx -k_2 A_2^2 \frac{\Delta f_{02}}{f_{02}} \tag{8.68}$$

$$V_2 = -\frac{1}{2} F_{02} A_2 \cos\phi_2 \quad \omega = \omega_0$$

Combining the three equations above:

$$\frac{\Delta f_{02}}{f_{02}} \approx -V_2 \frac{1}{k_2 A_2^2}$$

$$\frac{\Delta f_{02}}{f_{02}} \approx \frac{1}{2}\frac{F_{02}}{k_2 A_2} \cos\phi_2$$

$$\frac{1}{T}\oint \frac{\partial F_{ts}}{\partial \mathrm{d}} dt \approx -\frac{F_{02}}{A_2} \cos\phi_2 \quad \omega = \omega_0 \text{ and } A_2 \ll A_1 \tag{8.69}$$

Equation (8.69) provides the means to compute the average of the gradient of the nonlinear term F_{ts} via the drive force, amplitude, and phase of the second mode. Importantly, equation (8.69) is fully general for any oscillator where the conditions expressed there are shown. Expressions similar to the one in equation (8.69) have been simultaneously exploited [22] with the terms for the virials and energy dissipation [23–25] to extract material properties from models such as those shown in equation (6.11).

References

[1] Raman A 2014 Atomic force microscopy https://nanohub.org/resources/520

[2] Paulo Á S and García R 2001 Tip-surface forces, amplitude, and energy dissipation in amplitude-modulation (tapping mode) force microscopy *Phys. Rev.* **B 64** 193411

[3] Amo C 2019 *Microscopía De Fuerzas Bimodal Y No Resonante Para Medir Propiedades Físicas Y Químicas a Escala Nanométrica* (Madrid: Universidad Autónoma de Madrid)

[4] Albrecht T R, Grutter P, Horne D and Rugar D 1991 Frequency modulation detection using high-Q cantilevers for enhanced force microscope sensitivity *J. Appl. Phys.* **69** 668–73

[5] Giessibl F J 1997 Forces and frequency shifts in atomic-resolution dynamic-force microscopy *Phys. Rev.* **B 56** 16010

[6] Gramazio F, Lorenzoni M, Pérez-Murano F, Evangelio L and Fraxedas J 2018 Quantification of nanomechanical properties of surfaces by higher harmonic monitoring in amplitude modulated AFM imaging *Ultramicroscopy* **187** 20–5

[7] Stark R W and Heckl W M 2000 Fourier transformed atomic force microscopy: tapping mode atomic force microscopy beyond the Hookian approximation *Surf. Sci.* **457** 219–28

[8] Stark R W 2004 Spectroscopy of higher harmonics in dynamic atomic force microscopy *Nanotechnology* **15** 347

[9] Basak S and Raman A 2007 Dynamics of Tapping Mode Atomic Force Microscopy in Liquids: Theory and Experiments *Appl. Phys. Lett.* **91** 064107–9

[10] Preiner J, Tang J, Pastushenko V and Hinterdorfer P 2007 Higher Harmonic Atomic Force Microscopy: Imaging of Biological Membranes in Liquid *Phys. Rev. Lett.* **99** 046102–5

[11] Turner R D, Kirkham J, Devine D and Thomson N H 2009 Second Harmonic Atomic Force Microscopy of Living Staphylococcus Aureus Bacteria *Appl. Phys. Lett.* **94** 043901

[12] Giessibl F J 2003 Advances in atomic force microscopy *Rev. Mod. Phys.* **75** 949

[13] Goldstein H, Poole C and Safko J L 2001 *Classical Mechanics* (London: Pearson)

[14] Steidel R 1989 *An Introduction to Mechanical Vibrations* 3rd edn (New York: Wiley)

[15] Lozano J R and Garcia R 2009 Theory of phase spectroscopy in bimodal atomic force microscopy *Phys. Rev.* **B 79** 014110

[16] Santos S, Gadelrab K, Font J and Chiesa M 2013 Single-cycle atomic force microscope force reconstruction: resolving time-dependent interactions *New J. Phys.* **15** 083034

[17] Garcia R and Herruzo E T 2012 The emergence of multifrequency force microscopy *Nat. Nanotechnol.* **7** 217–26

[18] Lozano J R and Garcia R 2008 Theory of multifrequency atomic force microscopy *Phys. Rev. Lett.* **100** 76102–5

[19] Santos S 2014 Phase contrast and operation regimes in multifrequency atomic force microscopy *Appl. Phys. Lett.* **104** 143109

[20] Santos S, Lai C-Y, Olukan T and Chiesa M 2017 Multifrequency AFM: from origins to convergence *Nanoscale* **9** 5038–43

[21] Kawai S, Glatzel T, Koch S, Such B, Baratoff A and Meyer E 2009 Systematic achievement of improved atomic-scale contrast via bimodal dynamic force microscopy *Phys. Rev. Lett.* **103** 220801–4

[22] Martinez-Martin D, Herruzo E T, Dietz C, Gomez-Herrero J and Garcia R 2011 Noninvasive protein structural flexibility mapping by bimodal dynamic force microscopy *Phys. Rev. Lett.* **106** 198101

[23] Herruzo E T, Perrino A P and Garcia R 2014 Fast nanomechanical spectroscopy of soft matter *Nat. Commun.* **5** 3126

[24] Lai C-Y, Perri S, Santos S, Garcia R and Chiesa M 2016 Rapid quantitative chemical mapping of surfaces with sub-2 nm resolution *Nanoscale* **8** 9688–94

[25] Lai C-Y, Santos S and Chiesa M 2016 Systematic multidimensional quantification of nanoscale systems from bimodal atomic force microscopy data *ACS Nano* **10** 6265–72

www.ingramcontent.com/pod-product-compliance
Ingram Content Group UK Ltd.
Pitfield, Milton Keynes, MK11 3LW, UK
UKHW051849100926
8556IPUK00005B/544

9 780750 358071